Patrick Moore's
Practical Astronomy Series

Springer
London
Berlin
Heidelberg
New York
Hong Kong
Milan
Paris
Tokyo

Other titles in this series

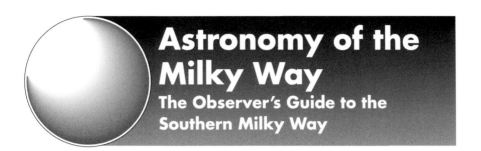

Astronomy of the Milky Way
The Observer's Guide to the Southern Milky Way

Mike Inglis

With 254 Figures
(including 21 in color)

Springer

Dr Michael D. Inglis, FRAS
State University of New York, USA

British Library Cataloguing in Publication Data
Inglis, Mike, 1954–
   Astronomy of the Milky Way
   The observer's guide to the southern Milky Way. – (Patrick
   Moore's practical astronomy series)
   1. Milky Way – Observers' manuals
   I. Title
   523.1′13
ISBN 1852337427

Library of Congress Cataloging-in-Publication Data
Inglis , Mike, 1954–
   Astronomy of the Milky Way/Mike Inglis.
      p. cm. – (Patrick Moore's practical astronomy series)
   Includes bibliographical references.
   Contents: pt. 1. The observer's guide to the northern Milky Way –
   pt. 2. The observer's guide to the southern Milky Way.
   ISBN 1–85233–709–5 (alk. paper) – ISBN 1–85233–709–5 (pt. 1) –
   ISBN 1–85233–742–7 (pt. 2)
      1. Milky Way. I. Title. II. Series.
QB981.I45 2003
523.1′13–dc21   2003050548

Patrick Moore's Practical Astronomy Series ISSN 1617–7185
ISBN 1–85233–742–7 Springer-Verlag London Berlin Heidelberg
a member of BertelsmannSpringer Science+Business Media GmbH
http://www.springer.co.uk

Typeset by EXPO Holdings, Malaysia
58/3830-543210   Printed on acid-free paper   SPIN 10926734

Dedicated to
the Astronomers of the Bayfordbury Observatory,
University of Hertfordshire,
past, present and future

Preface

Sometime during the last century when I was a boy, I remember looking at the night sky and being amazed at how bright and spectacular the Milky Way appeared as it passed through the constellation of Cygnus. It was an August Bank Holiday in the UK, and so, naturally, it was cold and clear. I may have looked at the Milky Way several times before that momentous evening but for some reason it seemed to stand out in a way it never had before. It was then that I began to observe the Milky Way as an astronomical entity in its own right and not as just a collection of constellations. It was also about that time that I had an idea for a book devoted to observing the Milky Way.

Fast forward a few years to a fortuitous meeting with John Watson, the astronomy editor of Springer-Verlag, who listened to my idea about a Milky Way book, and agreed that it would be a good idea. So I began, writing down notes and traveling the world, but at the same time observing hitherto uncharted regions of the sky (for me anyway!) and delighting in the new wonders it presented. After what seemed like an age, the book was completed, and you hold the finished product in your hands.

However, along the route I have been helped and guided by many people, both astronomers and nonastronomers, and I want to take this opportunity to thank them for taking part in what was a long-cherished labour of love. Firstly, my publisher, John Watson, whom I mentioned above and who has overseen the project from initial idea to completed book, and without whose help this book would never have seen the light of day. His knowledge of publishing and its many aspects is impressive. Add to this the fact that he is also an amateur astronomer himself, and you have a potent combination.

I have also been fortunate to have the company and friendship of amateur and professional astronomers worldwide, who freely gave advice and observational anecdotes that have subsequently appeared in the book. Amateur astronomers are a great bunch of people and none more so than Michael Hurrell and Don Tinkler, fellow members of the South Bayfordbury Astronomical Society. Their companionship has been a godsend, especially when life and its many problems seemed to be solely concerned with preventing me from ever finishing the book. Thank you, chaps!

I have also had the good fortune to be associated with many fine professional astronomers, and so I would like to especially thank Bob Forrest of the University of Hertfordshire Observatory at Bayfordbury for teaching me most of what I know about observational astronomy. Bob's knowledge of the techniques and application of all things observational is truly impressive, and it has been an honor to be at his side many times when he has been observing. Furthermore, I must mention Chris Kitchin, Iain Nicolson, Alan McCall and Lou Marsh, also from Bayfordbury, for not only teaching me astronomy and astrophysics, but for instilling in me a passion to share this knowledge with the rest of the world! I am privileged to have them all as friends.

Several nonastronomy colleagues have also made my day-to-day life great fun, with many unexpected adventures and Jolly Boys' Outings, and so it is only right and proper that the guilty be named – Bill, Pete, Andy and Stuart.

However, astronomy and the writing of books is, shall we say, only a meteor-sized concern, when compared to the cosmological importance of one's family. Without their support and love – especially when I was writing a book – patience and understanding, I would never have completed the project. Firstly, I must thank my partner and companion Karen, as we whiz together through space on our journey towards the constellation Hercules. Her patient acknowledgment that astronomers are strange people and that sometimes astronomy is *the* most important subject in the universe has made my life a wonder. At times, when it seemed as if I would never finish the book, and the road ahead looked bleak and cloud-covered, she would come into the study with a cup of tea, a Hobnob and a few gentle words of encouragement, and suddenly all was well with the world. Thank you, *Cariad*. Then there is my brother Bob. He is a good friend and a great brother and has been – amazingly, still is – supportive of all I have tried to achieve. Finally, I want to thank Mam, who has been with me all the way, even from before I saw the Milky Way in the garden in St Albans. She tells me that she always knew I would be an astronomer, and that it comes as no surprise to her to know that her son still spends a disproportionate amount of time standing outside in the cold and dark in the dead of winter and the middle of the night!

To all who have helped me become an astronomer and who make my life a lot of fun, many thanks, and don't forget the best is yet to come.

Dr Mike Inglis
St. Albans,
Hertfordshire, UK

and

Lawrenceville,
New Jersey, USA

August 2003

Arrangement of Book 2

During the course of writing this book, it became very apparent that it was going to be big – too big in fact to be totally contained within a single volume! This would have negated the premise of it being an observing guide that would and should be used at the telescope, not counting the cost of such a large book. After discussion with the publishers, it was decided to divide the book into two volumes: Book 2, which you now hold in your hands, would cover those constellations in the Milky Way that transit during the winter and spring months, from January to June, and are thus best placed (but not exclusively) for observation in the southern hemisphere; whereas the accompanying volume, Book 1, would cover those Milky Way constellations that transit during the summer and autumn months, and so are best seen from the northern hemisphere (but again but not exclusively).

However, experienced astronomers will know that a considerable number of Milky Way constellations residing in the southern part of the sky can be relatively easily seen from the northern hemisphere, and vice versa. In addition many constellations that straddle the celestial equator can be seen by both southern and northern hemisphere observers, and from an observational point of view this simply means that quite a significant amount of overlap can occur. It is therefore possible for an observer living in, say, Australia to make use of a considerable portion of Book 1 (which deals with the northern sky) and likewise for an observer in, say, the UK, to find Book 2 (covering the southern sky) quite useful.

In order to minimize repetition of data, and so reduce book size and cost, the duplication of information has been avoided where possible. However, to ensure that each book is self-contained, the introductory chapters, as well as the appendices and object index, are in each book.

Acknowledgments

I would like to thank the following people and organizations for their help and permission to quote their work and for the use of the data and software they provided:

- Michael Hurrell and Donald Tinkler of the South Bayfordbury Astronomical Society, England, for use of their observing notes.
- Robert Forrest, of the University of Hertfordshire Observatory at Bayfordbury, for many helpful discussions and practical tutorials over several years, on all matters observational.
- Dr Jim Collett, of the University of Hertfordshire, UK, for information pertaining to the Milky Way.
- Dr Stuart Young, formerly of the University of Hertfordshire, UK, for many informative discussions relating to the Milky Way.
- The astronomers at Princeton University, USA, for many helpful discussions on the Milky Way.
- The European Space Organization, for permission to use the *Hipparcos* and *Tycho* catalogues.
- Gary Walker, of the American Association of Variable Star Observers, for information on the many types of variable star.
- Cheryl Gundy, of the Space Telescope Science Institute, Baltimore, USA, for supplying astrophysical data on many of the objects discussed.
- Dr Chris Packham, of the University of Florida, for information relating to galaxies, particularly active galaxies.
- The Smithsonian Astrophysical Observatory, for providing data on many of the stars and star clusters.
- Richard Dibon-Smith of Toronto, Canada, for allowing me to quote freely the data from his books, *STARLIST 2000.0* and *The Flamsteed Collection*, and for the use of several of his computer programs.
- The Secretariat of the International Astronomical Union for information pertaining to the Milky Way.
- The publishers of *The SKY Level IV* astronomical software, Colorado, USA, for permission to publish the star charts.

I would also like to take this opportunity to thank several amateur astronomers who indulged me in order to discuss matters concerning the layout of the book. They are a great bunch of dedicated people: Dave Eagle (UK), Peter Grego (UK), Phil Harrington (USA), John McAnally (USA), Paul Money (UK), James Mullaney (USA), Wolfgang Steinicke (Germany), Don Tinkler (UK) and John Watson (UK).

I must also make a special mention of the astrophotographers who were kind enough to let me publish their images. They have taken the art and science of astrophotography to new heights, yet still remain awed and humbled by the beauty of the objects whose images they capture. In a time when the actual process of photography and CCD imaging seems to assume a higher importance than the very objects being imaged, this gifted and talented group of people remind us all that astronomy is a beautiful, unique and wondrous subject, and we would all do well to keep this in mind. They are (in alphabetical order): Matt BenDaniel (USA), Mario Cogo (Italy), Bert Katzung (USA), Dr Jens Lüdeman and members of the IAS (Germany), Axel Mellinger (Germany), Thor Olson (USA), SBAS (UK), Harald Strauss and members of the AAS (Austria), Chuck Vaughn (USA) and the Students for the Exploration and Development of Space (SEDS).

In developing a book of this type, which presents a considerable amount of detail, it is nearly impossible to avoid errors. If any arise, I apologize for the oversight, and I would be more than happy to hear from you. Also, if you feel that I have omitted a star or object of particular interest, again, please feel free to contact me at: astrobooks@earthlink.net. I can't promise to reply to all e-mails, but I will certainly read them.

Contents

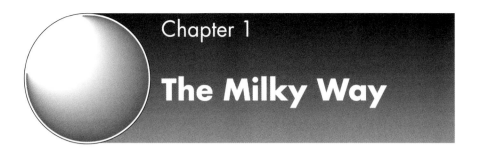

1.1 How to Use this Book

Most of us are familiar with the Milky Way. We may be lucky enough to live in a dark location and can see the misty band of light that stretches across the sky (see Figure 1.1). Others may live in an urban location and so can only glimpse the Milky way as a faint hazy patch that envelops several constellations. But how many of us make a point of observing the Milky Way as a celestial object in its own right? Very few I imagine, which is a pity as it holds a plethora of wonderful delights, ranging from deeply colored double and multiple star systems to immense glowing clouds of gas and mysterious dark nebulae which literally blacken the sky. It also holds quite a few star clusters that really do look like diamonds sprinkled on black velvet, not to mention the occasional neutron star, black hole and possible extra-solar planetary system! In fact, you could spend an entire career observing the Milky Way.

The Milky Way passes through many constellations; some are completely engulfed whilst others are barely brushed upon. Also it passes through both the northern and southern parts of the celestial sphere, making it a truly universal object, allowing it to be observed from anywhere in the world (see Star Chart 1).

The object of the book is not to give an introduction to the astrophysics of the Galaxy we live in, the Milky Way – there are many books listed in the appendices that can do that – but rather to introduce you to the many objects that can be observed that lie within the Milky Way. It may come as a surprise to you to know that you can observe the Milky Way on any clear night of the year, from any location on the Earth. So you could be observing from, say, Australia or Scotland. It wouldn't matter, as the Milky Way, or rather, particular parts of it, will be visible to you.

I have covered the complete Milky Way in this book, and so that means that there will be areas of it, and thus constellations, that may be unobservable from where you live. For instance, the constellation of Crux is a familiar one to observers living in Australia and New Zealand, but completely unobservable to European observers. Likewise, Camelopardalis and parts of Cepheus may be familiar friends to northern European observers, but are hidden from the view of our southern colleagues. What this does mean, however, is that this is a truly universal book that can be used by any astronomer anywhere in the world.

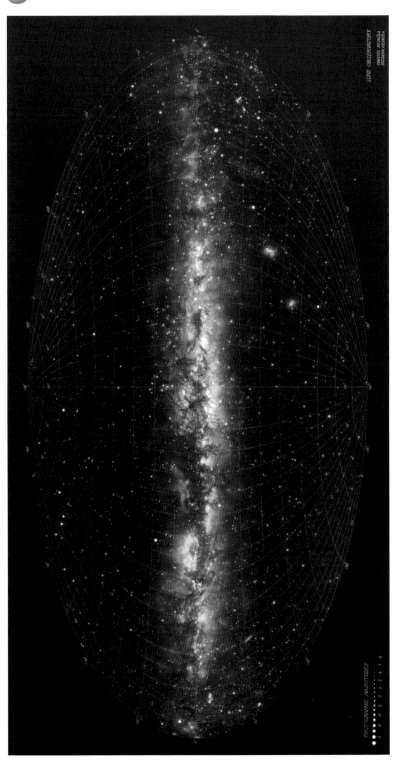

Figure 1.1. The Milky Way (The Lünd Observatory).

Many of the objects mentioned will, of course, need some sort of optical equipment, but a significant number are naked-eye objects, which is appropriate as the Milky Way itself is a naked-eye object, and the biggest one at that! But there are also quite a considerable number that only require small telescopes or binoculars, and by small I mean, say, 6–10 cm aperture. There are also those objects that will require a somewhat larger aperture, say, 10–25 cm, and the majority of the faint objects are in this aperture range. But not to exclude those observers with large telescopes I have also mentioned a few objects where very large apertures will be needed. Thus there is something for all amateur astronomers to see.

But remember that the one pre-eminent type of object that is just perfect to view with binoculars, and is truly impressive, are the many rich and awe-inspiring star fields or star clouds. This is what makes the Milky Way so spectacular. On a clear evening, one can observe Cygnus, or Vulpecula, or Sagittarius, or Centaurus, and literally be transported to other realms. The sights that fill the field of view cannot really be described, and once seen are never forgotten.

It goes without saying that a good star atlas is an essential part of every amateur astronomer's arsenal and fortunately there are many fine atlases to be had. A fine example of an atlas that is perfect for naked-eye observing is the redoubtable *Norton's Star Atlas*. Armed with this, and perhaps a pair of binoculars, you will have a lifetime of opportunity. For those who need a more detailed atlas, there are two that warrant attention: *Sky Atlas 2000.0* and *Uranometria 2000*. Both of these cover most, if not all, of the objects mentioned in the book, and will allow you to locate and find most of the fainter and not easily recognizable objects. It is also possible these days to have planetarium software on a computer and these too are fine tools to have, many allowing detailed star charts to be printed.

An astute observer will notice that the boundary of the Milky Way that I have adopted may not be the same as that in, say, *Norton's*, or older star atlases. I have in fact taken the boundary to be that identified by the Dutch astronomer Antonie Pannakoek, who measured the density of stars in the sky, and ascribed a limiting factor to the star density which enabled a boundary to be placed on the visible Milky Way. We are of course completely immersed in the Milky Way and most stars, nebulae and clusters that we observe are in fact within the Milky Way.[1] Thus the misty band of light that we call the Milky Way is just a region of the sky where the number of stars is so large so as to make a distinct and visible impression.[2]

However, there is a downside to adopting this boundary, as many of the favorite objects are left out if they do not lie within the Milky Way. Examples of such passed-over showpieces are the Pleiades and the Andromeda Galaxy, to name just a couple. I actually had to be quite strict in this respect, as if I were to include those objects that just lie outside of the Milky Way, the size and cost of the book would have doubled.

The layout of the book is straightforward. I have grouped the constellations that lie within the Milky Way more or less[3] in order of the month at which they transit at midnight. This means that the constellation will be at its highest point above your horizon at midnight. The reason for this is quite simple: if I were to describe in detail all the Milky Way constellations that can be seen at any one particular time of the year, not only would I be repeating a substantial amount of information, but the book would probably be about 900 pages long! Thus, for January and February, I discuss the Milky Way in Monoceros and Canis Major, to name but a few. However, seasoned amateurs will know that there are other constellations besides these that are in the Milky Way that can be seen during these

---

[1] There are a few objects that can be observed that are actually located outside of the Milky Way (and I don't mean galaxies!).
[2] The boundary I use is also the one adopted by the International Astronomical Union.
[3] There are exceptions to this, as described in the text.

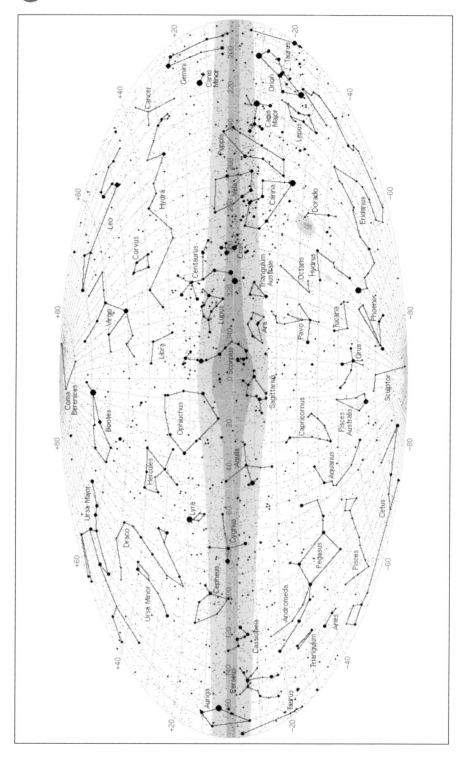

months, and this is perfectly true, but they do not transit at midnight during these months! The other constellations may be rising at midnight, or setting, or something in between, but they will not be at their best position for observation, *at midnight!* It is just a convenient means of presenting the data in a reasonable manner. You can of course view other parts of the sky during these months, say Cygnus, but it may not necessarily be at its optimum observing position. In fact, for the example given, it will be so low down as to be nearly unobservable. Nevertheless, it will be there for you to look at.[4]

For the sake of completeness, however, at the end of each chapter I have listed, for both northern and southern observers, those Milky Way constellations that are also visible but with the above caveat in mind! Armed with this knowledge, you can go out and observe quite a large proportion of the Milky Way on any clear night of the year, from anywhere in the world.

In addition, I do not structure each chapter in any formal way, but rather in a manner that seems appropriate. For instance, in Orion, I start off by describing the many wonderful double and multiple stars that the constellation has, whereas in Cassiopeia I begin with detailed descriptions of its many glorious star clusters.

Throughout the book are many simple star charts, and they are just that – simple! They are not meant to take the place of a star atlas, but are rather a pointer in the right direction. Also, I have had the opportunity to include quite a few wonderful photographs and CCD images of many of the objects described. These were taken by gifted and talented astrophotographers and astro-CCD imagers and to include them in the book is a privilege. You may notice, by their conspicuous absence, that there are no drawings of any of the objects in the book. The reason for this is simple. Not only can I not draw to save my life, but drawings or sketches, particularly of astronomical objects, are very personal constructs and more often than not do not resemble the generally recognized shape or form of an object. Rather, they describe what you, the observer, can see at that particular moment. It is no exaggeration to say that one can take two observers, show them the same object through two identical telescopes at the same time, and ask them to draw it, and you will end up with two quite different and distinct drawings. I believe it serves no useful purpose for me to include drawings of objects that show how I see a particular cluster or nebula, as it will be different from what you see. Furthermore, I agree with what the astronomer David Ratledge says in his book on the Caldwell objects, when he asks how can one really sketch something in detail when you are using averted vision?

Finally, at the end of each chapter is a list of the main objects mentioned in the text, giving their positions in right ascension and declination. This will allow you to use a star atlas, and the GOTO facility of your telescope or the setting circles on your telescope in order to locate and observe the objects. I also include, where appropriate, the objects' magnitude and, for double stars, their separation and position angles.

I have tried to include not only the well-known Messier, Caldwell and NGC objects that we are all familiar with, but also those that are perhaps less familiar to you. They may be faint and/or small, but they are all definitely worth observing. If I have left out an object that may be a particular favorite of yours, then I apologize, as I tried to include as much as I could.

---

[4] I had quite a detailed correspondence with several amateur astronomers from the UK, Australia and the USA about how to present the data, and this method was the one that most of them preferred.

◀ **Star Chart 1.** The Milky Way in Galactic Coordinates. Compare this star map, that shows the Milky Way superimposed over the constellations, with the images of the Milky Way at the end of the book. Note that the contours of the Milky Way are approximate. Star map courtesy of Richard Powell.

So, enough of the words, but before we begin a year-round voyage of the Milky Way, I would like to take this opportunity to discuss a topic that is central to the subject matter of this book, and that every astronomer should be aware of.

1.2 A Plea to the Faithful

We live in a world where science and especially astronomy is making great leaps forward in our knowledge and understanding of the Universe. Every day there is a news article on some new discovery, either from an Earthbound telescope or satellite in space, or a new image is published of some magnificent and mysterious object deep in outer space.

At the same time more and more people are becoming interested in amateur astronomy. Telescopes are getting cheaper, better, and packed with additional extras like thousand-object databases and equatorial mounts. And, of course, the Internet is a vast resource of information on everything astronomical.

But one thing worries me – the ever increasing plague that is light pollution and especially when it concerns the topic of this book – The Milky Way. How many of us can remember a time when we could go out into our gardens or a nearby park and see the wonderful swathe of the Milky Way cut a path across the sky? Nowadays one needs to be deep in a sparsely populated rural landscape or high in the lonely mountains in order to see this wonder of nature.

We are told constantly that the resources and animals of the world we live in need to be conserved and protected, and I agree wholeheartedly with this notion. I have never seen a blue whale, or an American bison, or even a monarch butterfly or a slipper orchid. Furthermore, I have never visited the Great Barrier Reef or the Brazilian rainforest; yet I feel strongly that they must be protected for all and for ever and I am not a biologist or ecologist. In the same vein, we should keep the seas clean, the landscapes natural and the atmosphere breathable. Yet, in all this, it seems to me that the conservation of nature and the appreciation of our world stops when it gets dark. Surely, the most wonderful spectacle in all of nature is the night sky, blazing forth in all its glory. Yet most of the world seems unaware that we are losing this resource. In a recent study published in the *Monthly Notices of the Royal Astronomical Society*, it stated: "about one-fifth of the world population, more than two-thirds of the US population and more than one-half of the EU population have already lost naked-eye visibility of the Milky Way … and about two-thirds of the world population and 99% of the population in the US (excluding Alaska and Hawaii) and the EU live in areas where the night sky is above the threshold set for polluted status".[5] This is a truly appalling statistic! And if you think I am a zealot and maybe overly dramatic, just think back to your own experience in this regard. How much of the night sky have you seen literally disappear in just a few years?

I would like to think that in the future it would be possible for me to take my children or grandchildren out into the garden and show them the Milky Way, and how splendid it looks, hoping that it inspires the awe and wonder in them that it did and still does in me. But if we do not try to curb the energy- and resource-wasteful spread of light pollution, this will not happen. But what can we do?

In order to try to come to an equitable balance between conservation and common sense, we need to be aware and appreciative to the wishes of the nonastronomer. It may be necessary to show them how beautiful, and, more importantly, how special the night sky and the Milky Way really are. Fortunately, there are many conservation societies through-

---

[5] P. Cinzano, F. Falchi, C.D. Elvidge. *Monthly Notices of the Royal Astronomical Society*, **328**, 689–707 (2001).

out the world that share this agenda, not forgetting the very important Dark Sky societies specifically aimed at reducing light pollution, and we as astronomers must promote their aims and agendas.

The night sky and the Milky Way are truly wonders of the Universe we reside in and are part of the place we call home, the Earth. They have been our companions since humans first looked up towards the stars many thousands and perhaps millions of years ago, and yet in just a few generations we may lose them. We need to, nay must, keep these wonders, not just for us, but for all people for all time. So please become aware that we are losing, slowly but surely, our access to the night sky. Join the conservation societies that actively promote safe and efficient night time lighting, and become an active member of the Dark Sky Association. Show your family and friends how amazing the night sky is and convince them that it must not be obscured any further.

As was once sung in a song, "we are star stuff", and we as astronomers, and in fact all people, would do well to remember this.

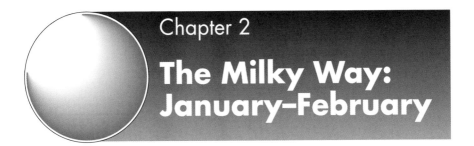

Chapter 2
The Milky Way: January–February

Monoceros, Canis Minor, Hydra, Canis Major, Puppis, Lepus, Columba, Pyxis, Antila, Vela.

R.A 6[h] to 11[h]; Dec. 12° to –28°; Galactic longitude[1] 200° to 290°; Star Chart 2, Figure 2.

2.1 Monoceros

The Milky Way at the start of a new year runs diagonally through the constellation **Monoceros** and in fact the galactic equator runs more or less straight along this path (see Star Chart 2.1). In the early part of January the constellation transits and is at its best for both northern and southern hemisphere observers. When we look out at Monoceros, we are actually peering deep into the **Orion Arm** of the Galaxy, in the opposite direction to the galactic center in Sagittarius. Many beginner observers are surprised to learn this, as the area is not ablaze with the star fields and dark dust clouds that one normally associates with the Milky Way. Indeed, at first glance the area seems quite empty and barren. However, this is a misconception as there are several open clusters, nebulae and stars that warrant our attention.

Probably the most famous, and certainly one of the most photographed objects (or, in these days, CCD imaged) is the area surrounding the star cluster **NGC 2244**, which lies about 2.5° east of Epsilon (ε) Monocerotis. The cluster itself contains about 100 stars and has a combined magnitude[2] of around 4.8, so that on a dark night it will appear to the naked eye as a fuzzy star.

When observed through a small telescope, the field is filled with a sprinkling of stars set against the faint glow of the Milky Way. The brightest stars in the cluster are two giant stars: a blue-white O5-type star and a distinctly yellow K0-type star,[3] known as **12 Monocerotis**.[4] A larger telescope of say 15 cm aperture will show dozens of stars with the brighter members forming nice paired arrangements. However, surrounding the cluster is

---

[1] See Appendix 1 for details on astronomical coordinate systems.
[2] See Appendix 2 on magnitudes for a description of integrated and combined magnitudes.
[3] See Appendix 3 for details on star type and classification.
[4] This star is in fact not part of the cluster, but a foreground star.

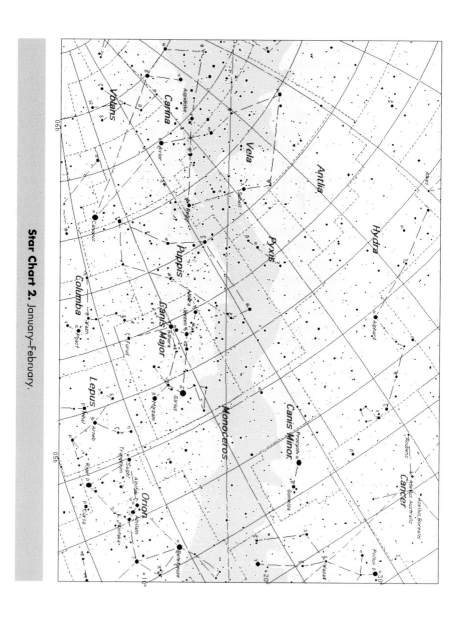

Star Chart 2. January–February.

Figure 2. The southern Milky Way. Note the enormous Vela SNR on the right of the image (Matt BenDaniel, http://starmatt.com).

an object that can, and should, impress the observer. It is the enormous cloud of glowing gas known as the **Rosette Nebula, NGC 2237–39 (Caldwell 49)** (see Figure 2.1). This is the left-over material out of which it is thought NGC 2244 formed. It lies at an estimated distance of about 1.74 kiloparsecs, and thus its apparent size of 80 by 60 arcminutes would imply that its actual size is an enormous 40 parsecs, making it one of the largest nebulae known.

This giant emission nebula has the dubious reputation of being very difficult to observe. But this is wrong – on clear nights it can be seen with binoculars as a soft glow surrounding the distinctive parallelogram of the cluster. The use of a nebula[5] (an [OIII] or Hβ filter) or light-pollution filter will help. It is over 1° in diameter, and thus covers an area of sky four times larger than a full moon! With a larger aperture and light filters the complexity

---

[5] See Appendix 4 for details of the various types of light filters available.

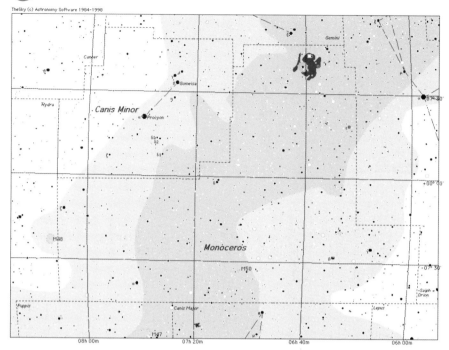

TheSky (c) Astronomy Software 1984-1998

Star Chart 2.1. Monoceros.

of the nebula becomes readily apparent, and under perfect seeing conditions dark dust lanes can be glimpsed. On rare occasions, dark and tiny circular patches can be seen in silhouette against the pale gray glow of the nebula. These are called Bok globules and are the precursors to star formation.[6] The brightest parts of the emission nebula have their own NGC numbers: **2237, 2238, 2239** and **2246**. It is a young nebula, perhaps only half a million years old, and star formation may still be occurring within it. Photographs show that the central area contains the star cluster NGC 2244, along with the "empty" cavity caused by the hot young stars blowing the dust and gas away. It is also known as the **Rosette Molecular Complex (RMC)**.

There is another patch of nebulae in Monoceros, which is equally as famous as the Rosette Nebula, but much harder to observer. The brightest part of nebulosity and stars is an open cluster known as **NGC 2264**, at magnitude 3.9 with about 40 stars.[7] What makes this otherwise unremarkable cluster special is that it has such a distinctive shape that it has been called the **Christmas Tree Cluster** (see Star Chart 2.2). Not only that, surrounding the cluster is a vast swathe of dark and bright nebulosity (its brighter section is also called NGC 2264 and illuminated by many of the cluster's type B stars; see Figure 2.2).

---

[6] The whole process of star formation would be a book in itself. I refer readers to the book *Observer's Guide to Stellar Evolution*, Springer-Verlag, for a complete and up-to-date simple introduction to the topic.

[7] The quoted magnitude of a cluster may be the result of only a few bright stars, or on the other hand may be the result of a large number of faint stars. Also, the diameter of a cluster is often misleading, as in most cases it has been calculated from photographic plates, which, as experienced amateurs will know, bear little resemblance to what is seen at the eyepiece.

Figure 2.1. Caldwell 49 (Chuck Vaughn).

Lying just 40 arcminutes to the south of the star cluster and seemingly floating in front of it is the famous **Cone Nebula**, one of the sky's justifiably finest dark nebulae. Believed to be at a distance of about 80 parsecs, the cone is at least 6 light years in length and is still an area of active star formation, with the cocoon of dust and gas surrounding several proto-stars. The dark nebula is about 5 arcminutes long and ranges from 40 arcseconds to 3 arcminutes wide. Photographs of this object are spectacular, which in one way is a shame as it is a very difficult object to observe and so gives a false impression of observability to an amateur. Nevertheless, with a large-aperture telescope of 40 cm, and using a nebula filter, the Cone may appear as a black smear set against gray background nebulosity. Lying at the base of the Christmas Tree Cluster is the star **S (15) Monocerotis**. This is a blue supergiant, type O7, and varies in magnitude from 4.2 to 4.6. It is also a visual double star, with a companion 2.8 arcseconds away, at magnitude 7.9.

The only Messier object in Monoceros and one that is often overlooked by amateurs is the fine open cluster **Messier 50 (NGC 2323)** (see Figure 2.3). Discovered by Cassini, this is a fine, heart-shaped cluster easily seen in binoculars, magnitude 5.9, and visible to the naked eye on clear nights.[8] Within the large, bright and irregular cluster of around 80 blue stars is a striking red star just 7 arcminutes south of the cluster's center. What makes the cluster particularly challenging is that the area of the sky where it resides is full of small stellar groupings and asterisms. The question often arises, where does the cluster end and the background star field begin?

---

[8] See Appendix 5 for details about star clusters.

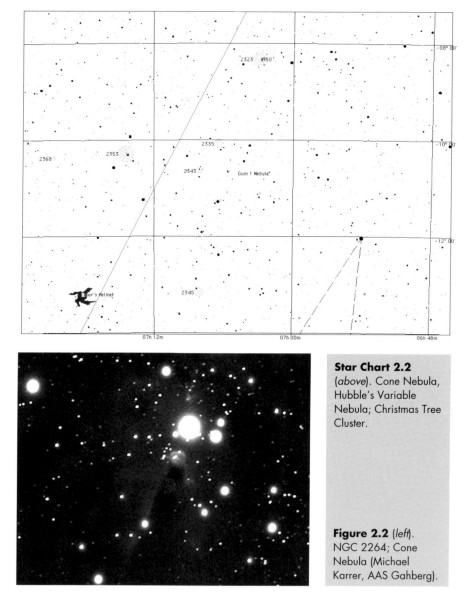

Star Chart 2.2 (*above*). Cone Nebula, Hubble's Variable Nebula; Christmas Tree Cluster.

Figure 2.2 (*left*). NGC 2264; Cone Nebula (Michael Karrer, AAS Gahberg).

There are several fine double[9] and multiple stars in the Milky Way sections of Monoceros, and one of the best is **Epsilon (ε) Monocerotis**. This is a lovely triple system of pale yellow stars along with a very faintly tinted blue companion,[10] magnitudes, 4.5, 6.5, and 5.6, seemingly set amongst the star fields of the Milky Way. Another is **Delta (δ) Monocerotis**. This is a nice blue-white star of magnitude 4.12, which has an unrelated companion **21 Monocerotis**, of magnitude 5.44. It is visible to the naked eye and easily

---

[9] See Appendix 6 for details about double stars.
[10] See Appendix 7 for details on the colors of stars.

Figure 2.3. Messier 50 (Harald Strauss, AAS Gahberg).

seen in binoculars. Perhaps the finest multiple star in this part of the Milky Way is **Beta (β) Monocerotis**. This is a magnificent triple star, first discovered in 1781 by Herschel. All the stars are a lovely steely blue-white in color. What makes this system so unique is that all the stars are very nearly equal in brightness. It lies about 11° south of the Rosette Nebula, and has a primary star, magnitude 4.7, with fainter companions of 5.2 and 6.1 magnitude.

A star that appears unremarkable, and is not mentioned too often, is **Plasketts' Star**. It is at magnitude 6.05 and is in fact a spectroscopic binary star system, type O8. However, measurements indicate that the system has a combined mass of over 110 times that of the Sun, and thus is one of the most massive of its type known. Telescopically, it is nothing special. However, it is sobering to know that when you observe this star, you are looking at one of the Galaxy's most exotic, albeit little known, objects.

Another, often overlooked open cluster is **NGC 2301** (see Figure 2.4). This is a very striking cluster measuring about 12 arcminutes in diameter, with a magnitude of around 6. In binoculars, a north–south chain of 8th and 9th magnitude stars is revealed, marked at its midway point by a faint haze of unresolvable stars. With a large aperture, there is a colorful trio of red, gold and blue stars at the cluster's center.

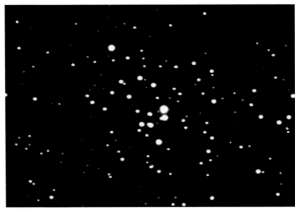

Figure 2.4. NGC 2301 (Harald Strauss, AAS Gahberg).

Figure 2.5. IC 2177. Photograph © Bert Katzung.

One interesting area of the Milky Way, which lies within a constellation that may be unknown to most amateurs, is the large region of stars and nebulosity that lies south of Messier 50. The reason for this is that the objects are best suited to photographic or CCD imaging, although some of the brighter parts can be seen in a 20 cm telescope. The nebula in question is **IC 2177 (Gum 1)** (see Figure 2.5), and its associated clusters, **NGC 2335, NGC 2343** (Figure 2.6), **Cr 465** and **Cr 466** (see Star Chart 2.3).

Figure 2.6. NGC 2343 (Harald Strauss, AAS Gahberg).

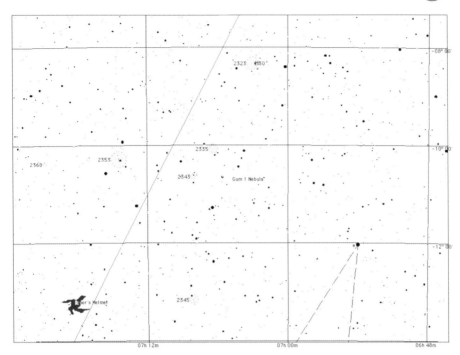

Star Chart 2.3. IC 2177 and surrounding clusters.

The emission nebula is in two parts: a 2° long north–south ribbon of gas and dust that in fact crosses into Canis Major, and a smaller nebula surrounding a star. Many astrophotographers and imagers liken the nebula to an eye with a pupil, as there is a horizontal dust lane through the nebula. It has been called the **Seagull Nebula** by some observers. The catalogue number IC 2177 designates only the area of seagull's wings. The "head" of the bird, **NGC 2327**, is the brightest portion of the nebula complex. It surrounds an 8th magnitude star and also includes some blue reflection nebulosity.

Our last look at open clusters in the Monoceros Milky Way is at **NGC 2353 (Herschel 34)** (see Figure 2.7) and **NGC 2506** (see Figure 2.8). The former lies a few degrees east of the nebulosity IC 2177. This is a small faint cluster best seen in a telescope and contains many orange 6th magnitude stars. It is a nice cluster of 30–40 stars and has a diameter of 20 arcminutes, with a combined magnitude of around 7.2 (see Star Chart 2.4). The latter cluster is a striking, but faint group of stars and is one of those passed-over clusters due to its low magnitude of 10.7. In an area only 7 arcminutes across, there are over 150 stars. Definitely worth seeking out!

It is only fitting that the final object we look at in Monoceros is also one of the most unique visible to the amateur. It is called **Hubble's Variable Nebula (Caldwell 46, NGC 2261)**, and is part of the aforementioned NGC 2264 nebular complex. It is easily seen in small telescopes of 10 cm aperture as a small, comet-like nebula, of size 2 × 1 arcminutes, which can be seen from the suburbs (see Figure 2.9).

What we are observing is the result of a very young and hot star clearing away the debris from which it was formed. The star **R Monocerotis** (buried within the nebula and thus invisible to us) emits material from its polar regions, and we see the north polar emissions,

Figure 2.7. NGC 2353 (Space Telescope Science Institute, AAO, UK–PPARC, ROE, National Geographic Society, and California Institute of Technology).

with the southern emission blocked from view by an accretion disk. It varies in brightness from 10th to 13th magnitude with an irregular period. What makes it especially interesting for us is that images of the nebula taken by amateurs several weeks apart will show a definite change in morphology. Some amateurs even claim that due to its high and unusual surface brightness, on extremely crisp nights with perfect seeing, it is possible, under high

Figure 2.8. NGC 2506 (Space Telescope Science Institute, AAO, UK–PPARC, ROE, National Geographic Society, and California Institute of Technology).

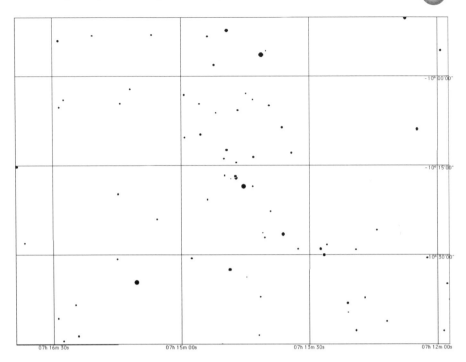

Star Chart 2.4. NGC 2353.

magnification, to draw the nebulosity, and see the changes in shape as compared to drawings made several months previously under similar conditions! The variability of the nebula, reported by Edwin Hubble in 1916, is due to a shadowing effect caused by clouds of dust drifting near the stars. It was also the first object to be officially photographed with the 200 inch Hale Telescope.

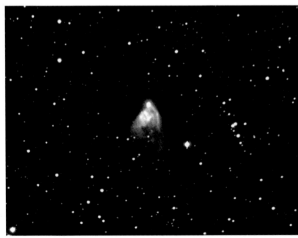

Figure 2.9. NGC 2261 (Harald Strauss, AAS Gahberg).

2.2 Canis Minor and Hydra

The Milky Way completely enfolds the constellation of **Canis Minor**, but only just manages to peek into **Hydra**. Nevertheless, there are only two objects that concern us here (see Star Chart 2.5).

In Canis Minor lies the star **Alpha (α) Canis Minoris**, better known as **Procyon**. What makes it interesting for us is that it is the eighth-brightest star in the sky, and is notable for the fact that it has, like nearby Sirius, a companion star which is a white dwarf star. However, unlike Sirius, the dwarf star is not easily visible in small amateur telescopes, having a magnitude of 10.8 and a mean separation of 5 arcseconds. It is magnitude 0.40, and type F5IV. It is also the fifteenth-nearest star, lying at a distance of 11.41 light years. The only justification for it being mentioned here is that it is a very easy object to observe.

In Hydra, however, is a much more interesting object visually: the open cluster **Messier 48 (NGC 2548)** (see Figure 2.10). This is a fine open cluster located on the outskirts of our Galaxy, hence its position in relation to the Milky Way – the edge! Located in a rather empty part of the constellation Hydra, about 4° southeast of **Zeta (ζ) Monocerotis**, this is believed to be the missing Messier object. It is a nice cluster in both binoculars and small telescopes. In the former, about a dozen stars are seen, with a pleasing triangular asterism at its center, while the latter will show a rather nice but large group of about 50 stars. Many amateurs often find the cluster difficult to locate for the reason mentioned above, but also because within a few degrees of M48 is another nameless, but brighter, cluster of stars which is often mistakenly identified as M48. Some observers claim that this nameless

Star Chart 2.5. Canis Minor; Hydra.

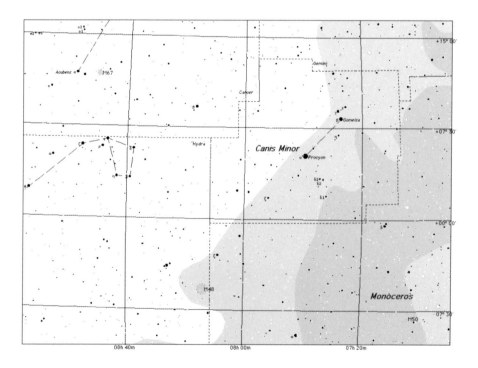

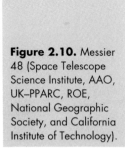

Figure 2.10. Messier 48 (Space Telescope Science Institute, AAO, UK–PPARC, ROE, National Geographic Society, and California Institute of Technology).

group of stars is in fact the correct missing Messier object, and not the one that now bears the name.

2.3 Canis Major

The Milky Way continues on its journey down into the southern sky by way of crossing the northeastern corner of **Canis Major** (see Star Chart 2.6). The constellation is below the

Star Chart 2.6. Canis Major.

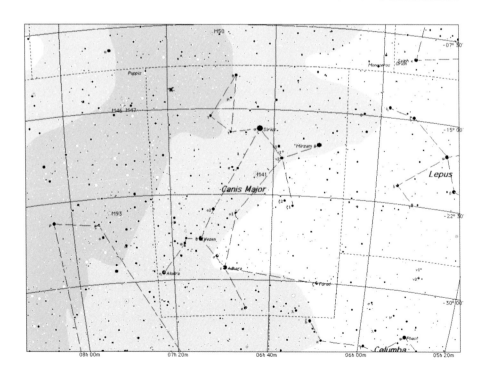

celestial equator so is never too well placed for northern observers; nevertheless it does present us with some fine objects to view. The constellation in its entirety is worth scanning with binoculars because although the bright Milky Way part of the constellation is easily seen, a large amount is actually swathed in obscuring dust which is a component of our Galaxy.

Let's begin by looking at the brightest star in the night sky – **Sirius**, or **Alpha (α) Canis Majoris**, also known as the **Dog Star**. It is the closest bright star visible from a latitude of 40° N, with a parallax of 0.3792 arcseconds. When observed from northerly latitudes, it is justly famous for the exotic range of colors it exhibits owing to the effects of the atmosphere. On the eve of the New Year, it will lie due south (or north in the southern hemisphere), and will culminate close to midnight.

It is not the most intrinsically luminous star in the sky, but appears bright because it is relatively close to us at a distance of only 2.7 parsecs, thus being the fifth closest star. It is an A1 V type star with a magnitude of –1.44 and a luminosity 23 times that of the Sun, i.e. 23 $L_{\odot}$. It also has a close companion star known as the **Pup (Sirius B)**, which is a white dwarf star, the first ever to be discovered.[11] However, trying to catch a glimpse of the Pup is notoriously difficult owing to its close proximity to Sirius itself. With a magnitude of 8.7 and a separation of only at best about 11 arcseconds, it presents a considerable observing challenge. It will be best placed during 2022, when it will be at its widest separation of 11.3 arcseconds and position angle 63°. It may be wise to put Sirius slightly out of the field of view, or even occult it in some manner in order to see the companion. Whichever method is used, the stars should be on the meridian, the seeing perfect, and you should have very clean optics. If you do glimpse it, you will be one of the few elite amateurs that have! Sirius is a dazzling sight in a telescope.

There are also some nice double stars visible in this part of the Milky Way, namely, **Pi (π) Canis Majoris** and **Mu (μ) Canis Majoris**. The former is an easily split system with a pale yellow primary and bluish secondary, although the secondary star is faint at magnitude 9.7, whilst the primary is at 4.7 magnitude. The latter system consists of two stars of differing brightness that nevertheless present a glorious double of orange and blue. The primary has a magnitude of 5.3, and the secondary 8.6.

There are a couple of triple star systems which are also visible: **17 Canis Majoris** is a nice triple system that exhibits a wonderful color contrast of white and orange-red stars. The stars are magnitude 5.8, 9.3, and 9.0 and should be split in binoculars and small telescopes. The other system **Tau (τ) Canis Majoris** is a gem! This is a wonderful example of its class. The triple is within the open cluster NGC 2362, and so its yellow and blue components are set against a glorious backdrop of faint stars, which are in turn set against the backdrop of the Milky Way. Bear in mind, however, that although the primary is at a magnitude of 4.4, the secondaries are faint at magnitudes of 10.5 and 11.2.

Open star clusters are in plentiful supply in Canis Major, but only a few of them reside within the confines of the Milky Way. These are Messier 41, Collinder 121, NGC 2354 and NGC 2362, and Caldwell 58.

Messier 41 (NGC 2287) is thought by many to be the finest cluster in the constellation and is easily visible to the naked eye on very clear nights as a cloudy spot slightly larger in size than the full moon, with a magnitude of about 4.5 (see Figure 2.11). Nicely resolved in binoculars, it becomes very impressive with medium aperture, with many double and multiple star combinations on display. There are about 70 members of the cluster and it contains blue B-type giant stars as well as several K-type giants. Current research indicates that the cluster is about 100 million years old and occupies a volume of space 80 light years in diameter.

---

[11] It was suspected by F. W. Bessel in 1834–5, but actually observed visually by A. G. Clark in 1862.

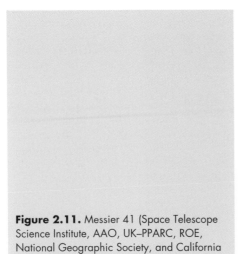

Figure 2.11. Messier 41 (Space Telescope Science Institute, AAO, UK–PPARC, ROE, National Geographic Society, and California Institute of Technology).

The cluster **Collinder 121** can present the observer with a problem, even though its combined magnitude is 2.8. It is a very large cluster, but one that is difficult to locate because of the plethora of stars in the background. At the northern border of the cluster is **Omicron (o) Canis Majoris**. It is best seen with large binoculars or low-power telescopes.

Open cluster **NGC 2354** is of a similar size to M41 and has about 70 stars. Because it is spread out over the sky, low powers are need here. On the other hand, **NGC 2362 (Caldwell 64)** is a delight (see Figure 2.12). At a magnitude of 4.1, this is a very nice cluster, tightly packed and easily seen, even with the smallest telescopes. With small binoculars the glare

Figure 2.12. NGC 2362 (Space Telescope Science Institute, AAO, UK–PPARC, ROE, National Geographic Society, and California Institute of Technology).

Figure 2.13. NGC 2360 (Space Telescope Science Institute, AAO, UK–PPARC, ROE, National Geographic Society, and California Institute of Technology).

from **Tau (τ) Canis Majoris** tends to overwhelm the majority of stars, although it itself is a nice star, with two bluish companion stars (recent research indicates that the star is a quadruple system). But the cluster becomes truly impressive with telescopic apertures; the bigger the aperture, the more stunning the vista. It is believed to be very young – only a couple of million years old – and thus has the distinction of being the youngest cluster in our Galaxy. It contains O- and B-type giant stars.

The final cluster, **Caldwell 58 (NGC 2360)**, is a beautiful open cluster, irregularly shaped and very rich (see Figure 2.13). There are many faint stars, however, so the cluster needs moderate-aperture telescopes for these to be resolved, although it will appear as a faint blur in binoculars, shining with a magnitude of 7.2. This is believed to be an old cluster with an estimated age of around 1.3 billion years.

Now for an observing challenge. Set against the backdrop of the Milky Way is a very faint and small planetary nebula – **IC 2165 (PK 221–12.1)**.[12] Shining with a magnitude of 12.9, it can, in principle, be glimpsed with a small telescope, say 10–12 cm aperture. But it is very small, with a diameter of about 8 arcseconds. These two factors make locating and identifying IC 2165 a problem. If low powers are used, it will resemble a star, and so high power is needed to resolve its nonstellar properties. With an aperture of 20 cm and more, the small, faintly blue disk can be seen. Using even larger apertures will resolve the noncircular shape along with a slight brightening at its center. Its central star has a magnitude of 15, and thus is very difficult to see (see Star Chart 2.7).

There are a few other nebulae in Canis Major, namely **NGC 2359 (Gum 4)** and an associated nebula **IC 468**. The emission nebula NGC 2359, also known as the **Duck Nebula**, is a moderately difficult object to find in small telescopes, but easier, naturally, in larger instruments, and so easily seen in telescopes of aperture 20 cm (see Figure 2.14). It consists of two patches of nebulosity, with the northern patch being the larger and less dense, the fainter being IC 468. With a telescope of aperture 12 cm, the nebula will look like a pale

---

[12] The PK designation refers to the planetary nebula catalog of Perek and Kohoutek (1967).

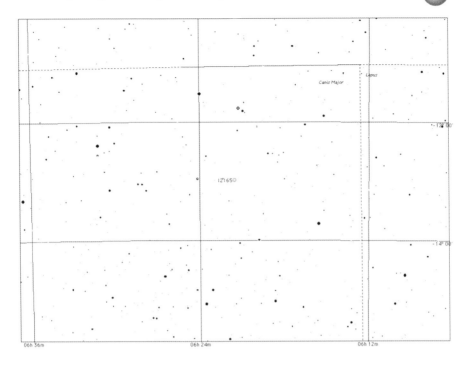

Canis Major

Lepus

IC 2165

-12° 00'

-14° 00'

06h 36m   06h 24m   06h 12m

Star Chart 2.7 (above). IC 2165.
Figure 2.14 (below). NGC 2359 (Georg Emrich and Klaus Eder, AAS Gahberg).

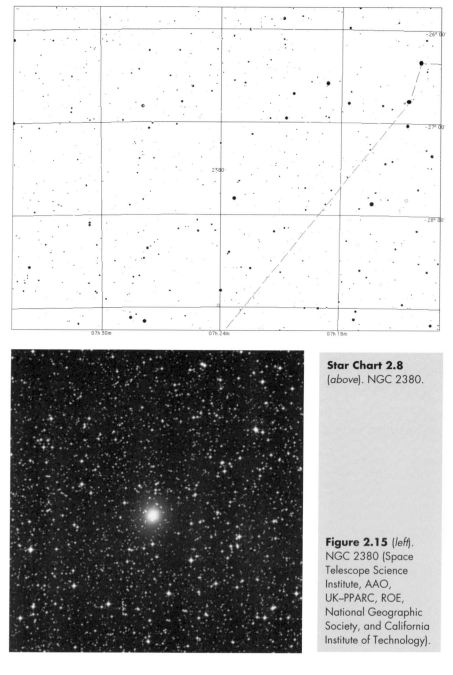

Star Chart 2.8
(*above*). NGC 2380.

Figure 2.15 (*left*).
NGC 2380 (Space
Telescope Science
Institute, AAO,
UK–PPARC, ROE,
National Geographic
Society, and California
Institute of Technology).

gray patch of light some 10 × 5 arcseconds in size. Telescopes with apertures of 40 cm and
more will show subtle distinctions in shade, but not much in detail, as the surface brightness is low. Under excellent conditions, IC 468 will appear as a ghostly ring of light some
20 arcseconds in diameter. Using an [OIII] filter will greatly improve the appearance of the
emission nebula, showing its delicate filamentary nature.

Even though we are looking into the Milky Way, some galaxies do appear, and one of these is the small spiral, **NGC 2380**. It is only 30 arcseconds across, but has a nice bright nucleus (see Figure 2.15 and Star Chart 2.8). Due to its position only 6° south of the galactic equator, a 20 cm telescope will be needed on order to observe it.

There are two additional nebulae I think make interesting objects. The first is **Sh 2–308**, which photographs show as a lovely bubble nebula surrounding the Wolf–Rayet star **EZ CMa**. Seen in a 30 cm telescope, only the brightest portion of the nebula can be observed, appearing as subtle brightening of the field, some 15 arcminutes west of the star. The use of an [OIII] filter will greatly enhance its visibility, so much so that a 20 cm telescope will now show some 20 arcseconds length of the nebula. The star EZ CMa shows only a small variation in magnitude. As mentioned earlier, there is also a cluster due south of the nebula, which includes Omicron^1 (o^1) Canis Majoris, called Collinder 121.

The other nice object is a remarkable reflection nebula surrounding the star **VY Canis Majoris**. The star, a very luminous supergiant, is a nice orange-red variable, is immersed within the reflection nebulae, and may well be an outlying member of the open cluster NGC 2362.

Finally, we look at a very odd variable star **UW Canis Majoris**. It lies only a short distance north of NGC 2362, and is an eclipsing binary star which is made up of two stars that rotate around each other in 4.4 days. What makes it so unusual from our point of view is that the stars are only about 0.2 astronomical units apart, and due to their high gravitational attraction to each other, they have distorted themselves into what are called flattened ellipsoids. Although the variations in magnitude are subtle at 4.7–5.0, and almost undetectable with the naked eye, it nevertheless is worthwhile seeking out for its "strangeness" factor!

2.4 Puppis

We now journey into those regions of the Milky Way that are, if not difficult to view, then completely hidden, for northern observers. The constellation **Puppis**, very low in the sky for observers in the UK, Europe and Canada, will be, conversely, high in the sky for observers in Australia, South Africa and South America. Nevertheless, there are a plethora of objects that can be seen wherever you live (see Star Chart 2.9)!

The constellation transits in mid-January, so of course this will be the best time to observe it. The galactic equator, and thus the Milky Way, run straight down the middle of it, and although there is no readily observable pattern or figure to the constellation, once found, it is never forgotten.

Many reference books say that there are as many as 70+ open star clusters in Puppis, but in reality many of these can be considered as just locally dense regions of the Milky Way. Other more conservative estimates put the number between 25 to 40 clusters, which seems much more realistic for small to medium aperture telescopes. To list and describe them all would entail a doubling (or tripling!) of the book's size and cost, so only the brighter will be mentioned. This shouldn't deter you, however, from sweeping the area with, say, binoculars or a rich-field telescope, as the views you will see of star-filled swathes of sky are wonderful. In fact, you could spend quite a considerable amount of time just scanning these rich star fields, such is their allure. Because we are looking out through the plane of the Galaxy, out to the galactic rim, dust clouds are few and far between, and this explains why we see so many clusters. On the other hand, due to the absence of gas and dust clouds, emission and reflection nebulae are absent for the most part.

Let's begin our observing schedule by looking in detail at some double stars.

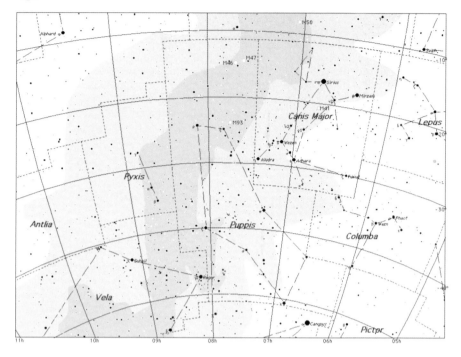

Star Chart 2.9. Puppis.

A nice pair of white stars set against a lovely star field is **Δ (Dunlop) 49**. Many of the background stars to this pair trace out long curving patterns. The pair are associated, but the period of their orbits seems to be extremely long as their separation changes only very slowly. **Sigma (σ) Puppis** is a brilliant deep orange giant star, type M0, of magnitude 3.3, with a giant white companion, type G5, at a faint magnitude of 9.4. Some observers believe the companion to be a yellow color, although this may be due to the contrast with the strongly colored primary. This pair is also set against the backdrop of a glittering star field. The primary is also a spectroscopic binary with a period of 257.8 days. A glorious double star that is located right in the middle of the Puppis Milky Way is **h (Herschel) 4046**. The primary is a bright orange, or golden, type G5 star, magnitude 6, with a white companion, type A, located about 22 arcseconds to the east, of magnitude 8.4. North of the companion star is a deep red star, and preceding the primary is a faint close pair. If a telescope of, say, 20 cm aperture is used the stars make a very attractive group set against the faint Milky Way backdrop. A colorful double, noted for its unusual pairing of a pale yellow primary and a red secondary is **h (Herschel) 4038**, easily visible and split by a telescope of aperture 7.5 cm. Another nice double is **Beta (β) 454**, consisting of a fine orange primary with a close white secondary. The background field is full of stars, and to the north of the pair lies a particularly prominent arc of stars. Our final double star is **H III 27**, a superb object for small telescopes. Both stars are white and nearly equal in brightness, set against a lovely star field. Some observers have claimed that the stars are pale yellow, but this may be due to their being observed whilst low down in the sky.

Let's now look at those objects for which Puppis is rightly famous: open star clusters.

Figure 2.16. Melotte 66 (Space Telescope Science Institute, AAO, UK–PPARC, ROE, National Geographic Society, and California Institute of Technology).

Our first cluster is thought to be one of the oldest open star clusters in the Galaxy, **Melotte 66**. It is a faint object, of magnitude 7.8, and telescopes of at least 20 cm will be needed to show it at all, and even then it is just a faint blob (see Figure 2.16). With a 30 cm aperture, a nice rich cluster becomes apparent with several brighter field stars scattered within and nearby to it. It is believed to be about 6 billion years old, and is around 7 arcseconds in diameter.

No constellation would be complete without its requisite Messier object, and Puppis is no exception, having three! Our first is **Messier 47 (NGC 2422)** (Figure 2.17). This is a wonderful object because it can be seen at all levels, from the naked eye – a faint blob – to a star-filled moderately rich cluster in the largest apertures. In binoculars it will be resolved into stars, and every increase in aperture will reveal more structure. It is a bright cluster with a magnitude of 4.4 and is about 30 arcseconds in diameter. The brighter members are blue-white B-type stars, but there is a nice contrasting orange variable star, **KQ Puppis**,

Figure 2.17. Messier 47 (Harald Strauss, AAS Gahberg).

Figure 2.18. Messier 46; NGC 2438 (Georg Emrich and Klaus Eder, AAS Gahberg).

located 40 arcseconds due west. With apertures of around 20–25 cm, two nice double stars can be seen. At the western edge of the cluster is **S 1120**, magnitudes 5.7 and 9.6, and in the center is **S 1121**, magnitudes 7 and 7.5.

Located just 1.5° away from M47 is the cluster **Messier 46 (NGC 2437)**. However, although, like M47, the cluster is large and bright, the resemblance ends there. M46 is a much more compact and richer cluster of medium-bright stars, and the difference can even be appreciated with binoculars (see Figure 2.18). As larger apertures are employed the number of stars visible increases dramatically. But what makes this cluster extra special is that a planetary nebula, **NGC 2438**, can be seen at the cluster's northeastern edge, glowing a pale bluish color (see Figure 2.18). The nebula does not actually reside within the cluster, but is a foreground object. To be seen in any detail, and not just as a fuzzy star, moderate magnification and apertures of at least 20 cm will be needed. Strangely enough, Messier did not report seeing the nebula. M46 is about 40 light years in diameter, and is believed to be ten times older, at 30,000,000 years, than M47.

Our last Messier object is another open cluster, **Messier 93 (NGC 2447)**. This has been described as a beautiful cluster, and deservedly so. Located about 1.25° northwest of the delightful yellow-golden star **Xi (ξ) Puppis**, the cluster is a bright object with fewer stars than M46. With a large aperture the view is stunning, with about 100 stars visible, along with a star-free gap in its center (see Figure 2.19). The field contains many doubles and triples, along with small groupings and star chains, and to the southwest can be seen two faint orange stars. All in all this is a very nice object.

Figure 2.19. Messier 93 (Space Telescope Science Institute, AAO, UK–PPARC, ROE, National Geographic Society, and California Institute of Technology).

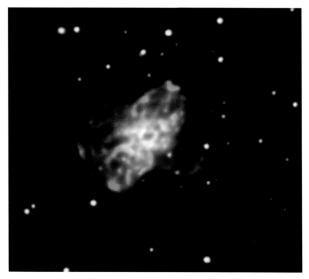

Figure 2.20. NGC 2440 (Space Telescope Science Institute, AAO, UK–PPARC, ROE, National Geographic Society, and California Institute of Technology).

There are two other planetary nebulae visible and worthy of our attention. One is **NGC 2440**, shining at a faint 11th magnitude. It is a bluish elliptically shaped object, some 15 by 30 arcseconds (see Figure 2.20). It can be seen in telescopes of 8 cm aperture, although just as a small blue blob (see Star Chart 2.10). The other is **NGC 2452**, which is fainter still, at 13th magnitude, and so something of a challenge to find in small telescopes (see Star Chart 2.11). It can be glimpsed with apertures of 20 cm, although there are reports of it being seen in a 10 cm telescope fitted with an [OIII] filter. It is about 20 arcseconds across, located in a small star field.

Let us return to open star clusters for a while and discuss three well-known ones. A fine object for binoculars and small telescopes is **NGC 2451**. It is a bright and large cluster centered on the K2-type star **c Pup**, magnitude 3.6, which is a fine orange color and so contrast nicely with the cluster stars (see Star Chart 2.12). With larger apertures several yellow and blue-colored stars become apparent. There are about 40 stars in the cluster.

Our next cluster is superb, and considered the finest object in the Caldwell catalogue. **NGC 2477** (**Caldwell 71**) is a beautiful cluster, very rich and concentrated (see Figure 2.21). Many of the stars form arcs with dark lanes running between them. With binoculars, the cluster is a faint haze, but increasing aperture will increase the spectacle until at a large aperture, say 30 cm, nearly 200 stars are visible; It is relatively old, with an age of about 1.2 billion years.

Finally, the cluster **NGC 2546** is worth observing. It is a very large cluster just south of the double star **h 4051**. Due to its size, a large field of view is necessary and so it is best seen in small telescopes (see Figure 2.22). A nice 5th magnitude star lies at its southern edge, whilst at the northwestern edge is a 6.4 magnitude bluish-white star. The cluster is a perfect object for small-aperture telescopes.

There is a globular cluster, **NGC 2298**, that is easily visible (see Figure 2.23). Shining at a magnitude of 9.4, it appears as a small, irregular object lying amongst a rich Milky Way backdrop. It can be seen easily enough with a telescope of, say, aperture 20 cm, but with very large aperture, resolution and structure becomes defined (see Star Chart 2.13).

Now for something rather special. Remember that we are looking into the plane of out Galaxy, which is crowded with dust and gas and stars and nebulae. This has the effect of blocking out light unless those objects are emitting light themselves. So it should come as no surprise to you to see that there is one class of object that has been missing so far –

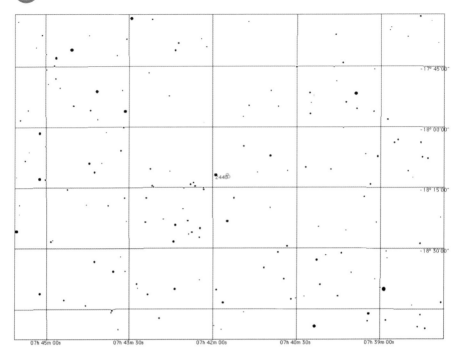

Star Chart 2.10 (*above*). NGC 2440.
Star Chart 2.11 (*below*). NGC 2452.

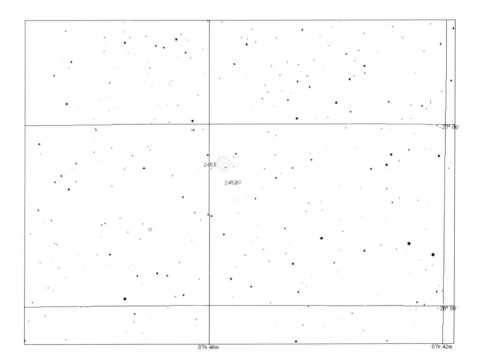

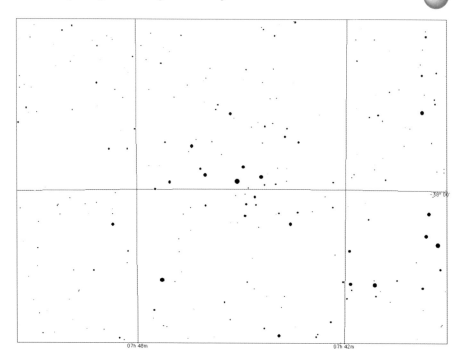

07h 48m   07h 42m

Star Chart 2.12. NGC 2451.

Figure 2.21. NGC 2477 (Space Telescope Science Institute, AAO, UK–PPARC, ROE, National Geographic Society, and California Institute of Technology).

Figure 2.22. NGC 2546 (Space Telescope Science Institute, AAO, UK–PPARC, ROE, National Geographic Society, and California Institute of Technology).

galaxies. We do not see many galaxies because the light from them cannot pass through the obscuring material. That's not to say there aren't any. Of course there are, but they are few and far between, and what's more usually very faint. However, there is one in Puppis, in the Milky Way part of the constellation, that we can observe, and that is **NGC 2525** (see Star Chart 2.14).

NGC 2525 is an 11th magnitude, barred spiral galaxy, located on the outer edges of the Milky Way. It will need a large-aperture telescope in order to be found, say 30 cm, and even then it will be faint and small (see Figure 2.24). The halo of the galaxy can be seen, although it is faint. At even larger apertures of around 40 cm, the halo and core can be resolved. A difficult object for sure, but a rare one in the Milky Way.

There is one variable star that merits a mention and that is L^2 **Puppis**. It lies in a small faint part of the Milky Way that seems detached from the main part. This of course is not the true state of affairs. Rather it is due to a large and very efficient dark cloud of dust that lies between us and the Milky Way, which is in fact part of a spiral arm, obscuring some of the light from the Milky Way, and so L^2 Puppis appears to be a separate and distinct object in its own right. The variable star in question is a semiregular, or long-period variable star, orange-red in color which varies in brightness over about 140 days. It ranges in magnitude from about 2.6 to 6.2, so is easily seen. Small telescopes equipped with simple spectrometers can easily see the dark lines and bands that the star produces.

Our last object in this busy constellation is the emission nebula **NGC 2467**, also known as **Gum 9**. Discovered by William Herschel in 1784, it is a small bright patch of haze embedded in a nice star field (see Figure 2.25). On one side of the nebula is a bright 8th magnitude star that may be responsible for emitting the energy that causes the nebula to glow. The central part has a very high surface brightness so that it can be easily located with large binoculars. Using small telescopes and averted vision, the nebula takes on a distinct elongated shape. The use of an [OIII] filter will greatly enhance the nebula, allowing its full extent to be appreciated.

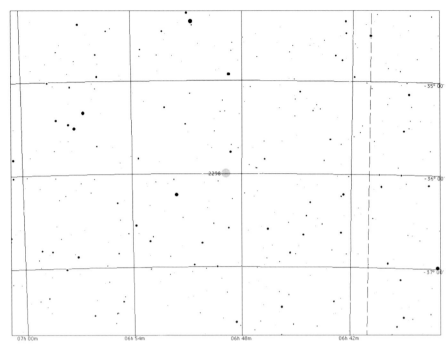

Star Chart 2.13
(*above*). NGC 2298.

Figure 2.23 (*right*).
NGC 2298 (Space
Telescope Science
Institute, AAO,
UK–PPARC, ROE,
National Geographic
Society, and California
Institute of Technology).

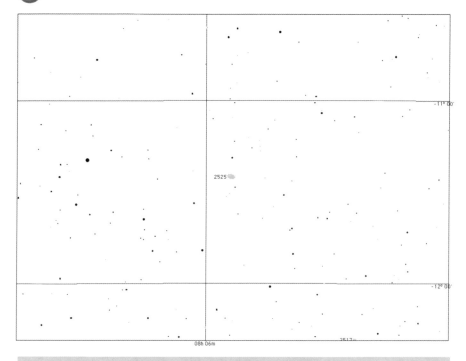

Star Chart 2.14. NGC 2525.

2.5 Lepus and Columba

The Milky Way extrudes[13] into the constellations of Lepus and Columba, albeit only for a very short distance (see Star Chart 2.15). I mention this only to be accurate, as there is

Figure 2.24. NGC 2525 (Harald Strauss, AAS Gahberg).

---

[13] A casual glance at a good star atlas will show immediately that this is the *only* word that can be used!

Figure 2.25. NGC 2467 (Space Telescope Science Institute, AAO, UK–PPARC, ROE, National Geographic Society, and California Institute of Technology).

absolutely nothing of interest in these Milky Way regions, other than for casual star-sweeping with binoculars, except for one multiple star, **h (Herschel) 3857**, which lies in the northeast corner of Columba. The star is a nice triple star system with two orange-yellow 5th and 9th magnitude stars, some 70 arcseconds apart, along with a 7th magnitude pale star which has been given the color ashy! It is easily visible in small telescopes and is set in a nice star field. Moving on.

2.6 Pyxis and Antila

If one looks in older star atlases, it will be apparent that the Milky Way only passes through the southwestern corner of **Pyxis** and barely reaches into the southern part of **Antila** (see Star Chart 2.16). In more recent and up-to-date atlases, a different scenario presents itself. Pyxis is almost completely engulfed, together with a substantial part of Antila. The latter is the approach I shall take here, as it allows more objects to be discussed.

The constellation Pyxis is only 221° square, with a handful of observably interesting objects in it. It transits at around early February, so this would be the best time to observe it.

A nice white star to begin with is **I 489**, which is in fact a very close double star. To all outward appearances it appears single in all but the largest telescopes of aperture 30 cm and greater, and may even appear slightly elongated as the two stars' images combine. It shines at magnitude 5.82. Another multiple star is **h (Herschel) 4166**, which may initially appear to telescope owners as a nice pair of white or pale yellow stars, but is in fact a triple system, although the fainter third component will remain unresolvable. Its two main members are at magnitudes 6.7 and 8.6. It lies in a very nice star field and so is ideal for small telescopes.

Strangely enough, in a constellation that is not particularly famous or stocked with stunning objects, Pyxis nevertheless does contain several galaxies, which are conspicuously

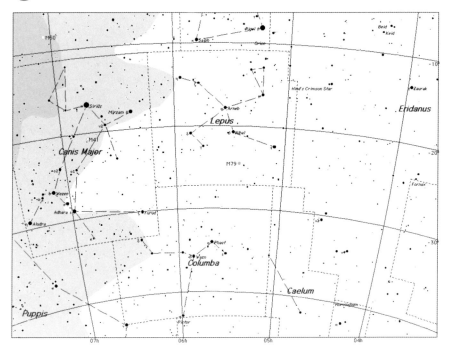

Star Chart 2.15 (*above*). Lepus; Columba.
Star Chart 2.16 (*below*). Pyxis; Antila.

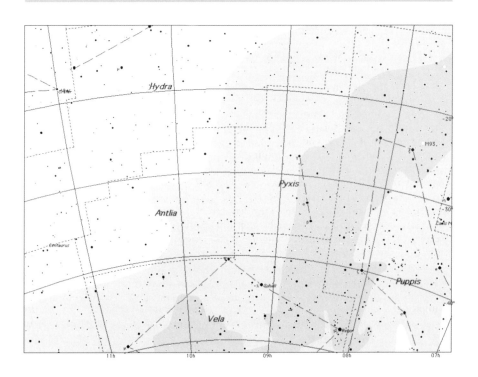

Figure 2.26. NGC 2613 (Space Telescope Science Institute, AAO, UK–PPARC, ROE, National Geographic Society, and California Institute of Technology).

absent from other Milky Way constellations. Because the Milky Way is less dense here, the galaxies' light can penetrate through the stars, dust and gas, which in those constellations that lie closer to the galactic equator, would be obscured. Saying that, bear in mind that there is still quite a lot of obscuration, which astronomers call reddening, and so the galaxies are still faint. Our first galaxy, **NGC 2613**, lies in a very nice star field and appears telescopically as an elongated spindle (see Figure 2.26).

The center is fairly bright whilst the ends of the spindle are very faint. It can be glimpsed with a 10 cm telescope, but a 20 cm aperture is really needed to study it in any detail. It has an integrated magnitude of about 10.5, and its dimensions are 7×2 arcminutes (see Star Chart 2.17).

Another galaxy well within the range of commonplace amateur telescopes is **NGC 2888**. This is somewhat fainter than the previous galaxy, having a magnitude of 12.3, and in all but the largest telescopes it will appear as a faint and very small object. There are several other galaxies that are visible, but only to those lucky individuals that have telescopes of aperture 40 cm and greater and have access to dark skies.

Another unique object is the combination of an open star cluster and a planetary nebula. This is **NGC 2818** and **NGC 2818A** (**PK 261+8.1**) respectively. The cluster is quite faint at about magnitude 8.2 and has around 40 to 50 stars in an area of about 7 arcminutes across, when seen through moderate telescopes (see Figure 2.27). The nebula lies in its western region and appears as a pale gray disk about 30–40 arcseconds in diameter. Using a higher aperture, some 80 stars can be seen and the planetary nebula takes on a distinct greenish color. Taken together this is a nice combination, but you will need at least a telescope of 15 cm to see them clearly (see Star Chart 2.18).

A final point is that on the later, more up-to-date star atlases, Pyxis is shown to be completely within the Milky Way, but at the eastern edge of the constellation is a large oval region where the Milky Way is not visible. What is happening here is that a very large amount of obscuration is taking place, blocking the light from the Milky Way. To my knowledge this has never been photographed by an amateur and so could provide an interesting project for someone.

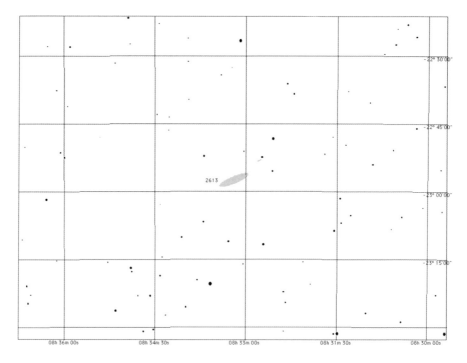

Star Chart 2.17. NGC 2613.

The constellation **Antila**, which transits at the end of February, lies on the outer regions of the Milky Way, yet perversely does not contain those objects we normally associate with the Milky Way, namely emission nebulae, and open star clusters. Instead, it has a scattering of galaxies, which alas show little difference from each other, and none are particularly bright.

The brightest galaxy is **NGC 2997**, which lies in a fairly scattered field of stars (see Star Chart 2.19). In small telescopes it will appear as a faint elliptical haze about 5 by 4 arcminutes in size, with a surprisingly bright nucleus about 10 arcseconds in length. It is at magnitude 9.3, and a telescope of, say, 15 cm will only just show it (see Figure 2.28).

However, increasing aperture and magnification will eventually show a surprising amount of detail so that in the largest telescopes of, say, 40 cm, a faint spiral structure can be see,[14] but only when using averted vision and the seeing is perfect.

2.7 Vela

The southern regions of the Milky Way contain many spectacular objects and vistas located in many splendid constellations, and many observers consider the constellation

---

[14] Some observers state that the spiral arms show a distinct coiled counter-clockwise spiral structure, similar to that seen in images of the galaxy.

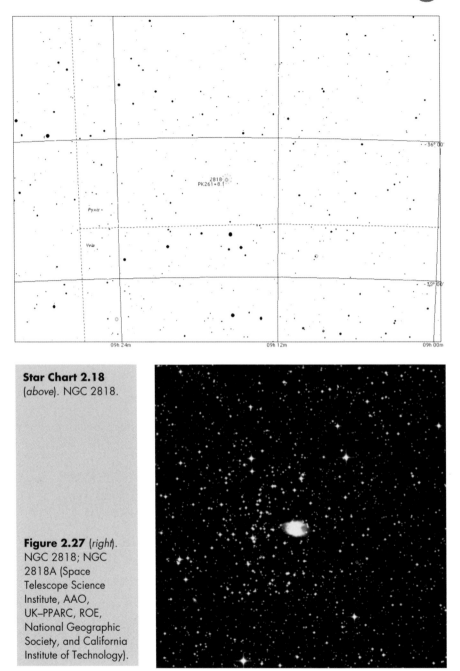

Star Chart 2.18 (*above*). NGC 2818.

Figure 2.27 (*right*). NGC 2818; NGC 2818A (Space Telescope Science Institute, AAO, UK–PPARC, ROE, National Geographic Society, and California Institute of Technology).

Vela to be one of the best (see Star Chart 2.20). Lying north of Carina, nestled between Puppis and Centaurus, and transiting towards the end of February, it is completely immersed in the Milky Way. Alas, for northern observers of latitude, say, comparable to the southern UK, it is forever below the horizon. From somewhere like New York or

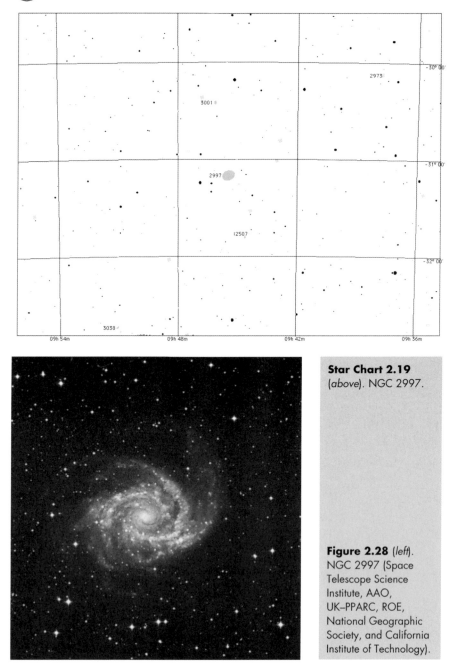

Star Chart 2.19 (above). NGC 2997.

Figure 2.28 (left). NGC 2997 (Space Telescope Science Institute, AAO, UK–PPARC, ROE, National Geographic Society, and California Institute of Technology).

Philadelphia, it skirts the southern horizon. It rises to an acceptable level for observers in Florida, but the true beauty of the constellation can only really be appreciated from such locations as Australia, New Zealand or South Africa.

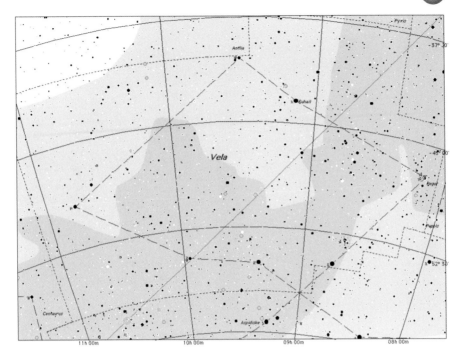

Star Chart 2.20. Vela.

The Milky Way here is rich in clusters although not all of them are suitable objects for medium-aperture telescopes. It also contains many attractive double and triple stars, one globular cluster, some superb examples of emission and reflection nebulae, and one spectacular supernova remnant. The constellation also contains many large and irregular patches of dark material, which obscure and dim many objects. However, this is particularly thin in the northeastern corner, which allows the light from several galaxies to peek through. Let's begin with some stars.

Our first star is **Gamma (γ) Velorum**, This is the brightest star in Vela, and is a nice double star. The primary **Gamma² (γ²) Velorum** is a brilliant white in color. Shining at magnitude 1.75, it contrasts nicely with its greenish-white, 4.27 magnitude, secondary **Gamma¹ (γ¹) Velorum**, separated by about 40 arcseconds. Just 1 arcminute southeast is a nice small double star system of white stars. What makes $γ^2$ Velorum special, however, is the interesting fact that it is the brightest example of a Wolf–Rayet star, with a spectral classification WC8. These are believed to be the precursors to the formation of planetary nebulae and are extremely luminous stars. Many of them have luminosities that reach 100,000 times that of the Sun with temperatures in excess of 50,000 K. Some measurements also indicate that $γ^2$ Velorum may also be the closest example of a Wolf–Rayet star to the Solar System. The second-brightest star in Vela is **Delta (δ) Velorum**. This is also a nice double of white stars, although many amateurs report that the 1.99 magnitude primary is yellow, whilst the 5.57 secondary is whiteish yellow. Separated by about 22.6 arcseconds, they make a fine sight in small telescopes. A fine triple system is **h (Herschel) 4104**, although in reality it is really a double star system with a close, but fainter, star nearby. The primary and secondary are both whiteish yellow in color, magnitudes 5.48 and 7.31, respectively. The primary is itself an extremely close binary star with a separation of only

Figure 2.29. NGC 2547 (Space Telescope Science Institute, AAO, UK–PPARC, ROE, National Geographic Society, and California Institute of Technology).

0.1 arcseconds. The triple is set against a lovely star field full of seemingly faint double and triple stars.

One of the classic double stars in Vela is Δ (**Dunlop**) **70**. The primary is a deep golden color, magnitude 5.14, and the pale yellow secondary shines at magnitude 7.05. They are separated by about 4.5 arcseconds. A nice object for small telescopes, it appears that the pair are physically associated. Furthermore, part of the Vela supernova remnant crosses the field of view, but be warned, this is only visible in large-aperture telescopes and even then will probably need some sort of [OIII] filter. Some double stars have a distinct color contrast, and such an example is Δ (**Dunlop**) **81**. This consists of a bright pair of 6th and 8th magnitude stars, but what makes it special is that the primary is yellow and the secondary is blue. Set against the faint haze of the Milky Way, this is a delight to observe. A double star that is often left off observing lists is **h** (**Herschel**) **4245**. This is a nice yellowish-orange and pale blue system with magnitudes 6.8 and 9.4 separated by about 9 arcseconds.

Set in a rich star field is the exquisite double star **Rmk 13 (I^A** and **I^B Velorum**). At magnitudes 4.48 and 8.36, these shine like small glittering diamonds, and are a splendid sight in telescopes of, say, 10 cm aperture. Our final double star that we will look at is **Mu (μ) Velorum**.[15] This is a nice system of yellowish stars, magnitudes 2.9 and 5.2 with a separation of about 2.5 arcseconds. The system is actually widening and so should become easier to observe. As a challenge, try observing **Lambda (λ) 108**, as this is a quadruple system with components at 2nd, 7th, 9th and 10th magnitudes.

As I mentioned earlier, there is a plethora of clusters in Vela, and to sweep the sky with binoculars or a rich-field telescope is reward enough, but let's look in detail at a few bright clusters. First there is **NGC 2547**, which is so large, some 17–20 arcminutes in diameter, that a large field of view is needed to encompass it in its entirety (see Figure 2.29). Therefore a low magnification is needed. Set against the backdrop of the faint Milky Way

[15] There are of course many more double, triple and multiple stars in Vela, but only the brightest are mentioned here.

Figure 2.30. IC 2391 (Space Telescope Science Institute, AAO, UK–PPARC, ROE, National Geographic Society, and California Institute of Technology).

there are several loops, chains and arcs of stars as well as doubles and groups of stars. With a magnitude of about 4.7 it should be visible to the naked eye. Some photographs and CCD images show the cluster to be enveloped in a very faint reflection nebula, although this has never, to my knowledge, been seen visually.[16]

A somewhat smaller and less rich cluster is **IC 2391** (**Caldwell 85**), containing about 25–30 stars (see Figure 2.30). It is a bright cluster though (the brightest in Vela), and contains the nice 3.6 magnitude star **Omicron (o) Velorum**, which is itself a spectroscopic binary. At an integrated magnitude of 2.5, and with a diameter of about 1°, it a great for binoculars and even appears as a faint fuzzy glow to the naked eye.

A fainter cluster is **IC 2488**, which lies 1° west of **N Velorum**. Although this cluster has an integrated diameter of about 7.4, most of its 80 members are of magnitude 10 or fainter, and so a telescope of aperture 25 cm is needed to fully resolve it (see Figure 2.31). Smaller telescopes will just show a faint haze with barely resolved stars. Going back to naked-eye clusters we have **NGC 3228** (see Figure 2.32). Visible as a 6th magnitude patch of light, it will improve when larger apertures are used, from about 15 stars set in a group some 18 arcseconds in diameter, to many more when viewed through, say, telescopes of 20 cm.

Sometimes one object can be mistaken for another, and this is certainly the case with **NGC 2660**. This is a distant open cluster that bears an uncanny resemblance to a globular cluster, some 11.5 arcseconds across (see Figure 2.33). It is a tight group of stars set against a nice rich star field with a magnitude of 8.8.

---

[16] I await the inevitable email telling me that it has!

Figure 2.31. IC 2488 (Space Telescope Science Institute, AAO, UK–PPARC, ROE, National Geographic Society, and California Institute of Technology).

Not all clusters managed to get into the NGC catalogue, for one reason or another, and such a one is the nice binocular object **Trumpler 10**. Maybe due to its absence from the catalog it is not too well known amongst amateurs, but it is a fine binocular object with

Figure 2.32. NGC 3228 (Space Telescope Science Institute, AAO, UK–PPARC, ROE, National Geographic Society, and California Institute of Technology).

Figure 2.33. NGC 2660 (Space Telescope Science Institute, AAO, UK–PPARC, ROE, National Geographic Society, and California Institute of Technology).

around 20 stars (see Figure 2.34). Seen in a small telescope it appears to have many more unconnected members set against the Milky Way backdrop.[17]

Some 10° north of the cluster IC 2391 (mentioned earlier) lies the rich, but small cluster **IC 2395**. It only has a small number of stars, about 40 in all, encompassed within a diameter of around 8 arcminutes. However, it is one of those objects that gets better and better when viewed with larger and larger telescopes. To the naked eye it appears as a fuzzy glow, but in binoculars and telescopes its true nature becomes apparent. An increase in aperture will show it as a splendid object with many more faint stars set against the rich background of the Milky Way. It really is one of those objects that warrants the description "a collection of small diamonds set upon black velvet".

Nebulae are objects Vela is not short of. In fact it has several superb examples and many of them are associated with the enormous, and famous, **Gum Nebula**. This nebula was discovered by the Australian astronomer Colin Gum, when he was making a mosaic of the region from wide-angle Hα photographs of Vela and Puppis. It is immense in size, over 40° × 30°, and can be thought of as the largest nebula in the sky (see Figure 2.35). Near its center is the Vela supernova remnant, or SNR, which color photographs show as a remarkably complicated network of filaments with a diameter of nearly 8°. Within the SNR is one of the very few pulsars that have been optically identified as being associated with an actual SNR. Let's look at a few of the nebulae present in Vela.

The **Vela Supernova Remnant** (SNR) is one of the few SNRs that are reportedly visible to the naked eye. Its filaments can be traced out over 5° of sky in declination, and one part, presumably the brightest, even has its own NGC number, NGC 2736 (see below). However, in order to see the nebula, the use of an [OIII] filter is needed – in fact, it is a necessity. You will, however, need exceptionally clear nights with superb seeing.

The one part of the Vela SNR that can be reasonably observed is **NGC 2736**. Using an [OIII] filter, it will appear as a longish and narrow pale streak of nebulosity set against a

---

[17] At the north of Trumpler 10 is another faint cluster, **NGC 2671**, as can be seen on the image.

Figure 2.34. Trumpler 10 (Space Telescope Science Institute, AAO, UK–PPARC, ROE, National Geographic Society, and California Institute of Technology).

Figure 2.35. Gum Nebula (Anglo–Australian Observatory/Royal Observatory, Edinburgh

rich star field,[18] it can be seen to stretch for about 20 arcminutes in length, but never more than 0.6 arcminutes wide. It varies in brightness along its length and seems more clearly defined along its western edge. Telescopes of aperture 20 cm and more will show it as a faint object, seemingly broken at many places along its length. Trying to observe this unique object can be frustrating, as a very clear night will be a prerequisite. In fact, its appearance or absence can be taken as a means of determining the local observing conditions.

Another emission nebula worth seeking out is **RCW 38**, also known as **Gum 23**. It can be glimpsed in a 20 cm telescope and its brightest part appears as a small faint irregularly shaped patch some 2 arcminutes in length, lying some 16 arcminutes south of the bright star **FZ Velorum**. If the conditions are suitable, then the nebula will appear to be split into two parts that have a faint tail extending to the north. Armed with an [OIII] filter, even more detail becomes visible, as more very faint nebulosity can be seen some 10 arcminutes southeast and 15 arcminutes northeast of the abovementioned bright part. Photographs show a large, wedge-shaped nebula that is split by several dark lanes of dust.

A further nebula, **RCW 40** (**Gum 25**), appears as a faint and indistinct hazy patch about 5 arcminutes across. Located within the nebulosity are five very faint stars of magnitude 11 to 12, arranged in a line. Its southeastern edge is more defined, and using a 20 cm telescope, along with the now ubiquitous [OIII] filter, it can be easily observed. Located some 25 arcminutes southeast of the nebula is the very compact star cluster **Markarian 18.**

Several planetary nebulae can also be found in Vela, including one that is thought to be amongst the finest in the entire sky. One of these is **NGC 2792**, a small gray-colored planetary nebula some 10 arcseconds in diameter. Its edges are well defined and show a plain smooth appearance. A small telescope of about 10 cm will show it, albeit as a tiny disk, but larger apertures will show it with ease (see Star Chart 2.21).

The next object is interesting: it is a planetary nebula, **NGC 2899**, which can be easily found with a moderately sized telescope, but located within the field of view is another recently located planetary nebula, **VBRC 2**. The nebula NGC 2899 is an irregularly shaped slightly elliptical object about 1.5 arcminutes in diameter. It is not uniformly bright, and needs care to be located although it can be found (eventually!) in a 15 cm telescope. Photographs show a distinct asymmetrical butterfly shape, which can just be glimpsed (or perhaps imagined?) when an [OIII] filter is used. However, in 1973, a very faint planetary nebula was discovered in the same low-magnification field of view as NGC 2899. This is VBRC 2 which is really a test for large telescopes of 30 cm and bigger. It appears as a faint round hazy patch about 2 arcminutes across. To aid recognition, some faint stars forming a triangle are superimposed upon it. You will need an [OIII] filter in order to observe and even locate this, as it will be invisible without it.

One of Vela's most famous objects is the planetary nebula **NGC 3132**, shining at a magnitude of 8.2, which makes it one of the brightest of its kind in the entire sky (see Figure 2.36). It can be easily located, as it lies 2.5° northwest of the 3.85 magnitude star **q Velorum**. Viewing the nebula through even the smallest telescopes, say, of 5 cm aperture will show a small hazy disk of gray light surrounding the 10th magnitude central star. With larger telescopes of, say, 15 cm aperture, and using a high magnification, then a bright ring of nebulosity will be seen with some parts appearing brighter than others. Its true diameter is some 40 arcseconds across. Photographs of it show a complex morphology, consisting of several oval rings all at differing angles, indicating that the star has undergone many outbursts of material. This has led to its name: the **Eight-Burst Nebula**. Surprisingly, the 10th magnitude central star is not the nebula's precursor star or the star that provides the radiation that illuminates the nebula. The true exciting star (the star that causes the nebula to

---

[18] Bear in mind that using an [OIII] filter, or any interference filter, will dim the light from stars, and so a rich star field may not appear so when viewed through the filter.

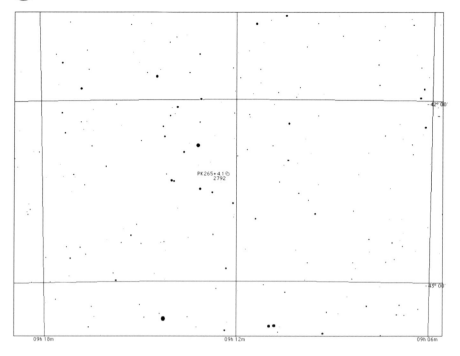

Star Chart 2.21 (*above*). NGC 2792.
Star Chart 2.22 (*below*). NGC 3132.

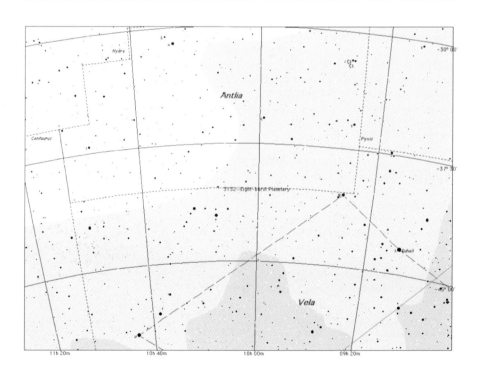

Figure 2.36. NGC 3132 (Space Telescope Science Institute, AAO, UK–PPARC, ROE, National Geographic Society, and California Institute of Technology).

glow) was discovered in 1976 and is the 16th magnitude bluish star some 1.7 arcseconds away (see Star Chart 2.22).

Our final objects include a reflection nebula, a globular cluster and surprisingly, given the amount of gas and dust present in this part of the Milky Way, a galaxy.

The reflection nebula **NGC 2626** is easily visible located around a 10th magnitude star (see Figure 2.37). With a diameter of around 2 arcminutes it is perceptibly brighter and

Figure 2.37. NGC 2626 (Space Telescope Science Institute, AAO, UK–PPARC, ROE, National Geographic Society, and California Institute of Technology).

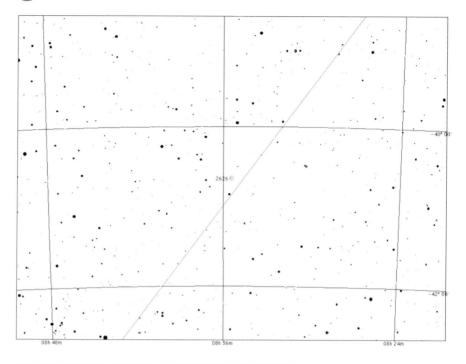

Star Chart 2.23. NGC 2626.

extended on its southern side. A 15 cm telescope will show it well providing the sky is dark (see Star Chart 2.23). Recent images indicate that it is located within the southern part of a much larger and fainter emission nebula called **RCW 27**.

The little-known globular cluster **NGC 3201 (Caldwell 79)** is one of the less condensed types of cluster (see Figure 2.38). With a magnitude of 6.8, it is a large and bright object but due to its low declination it is almost unknown to northern observers. In a 15 cm telescope it appears as a 5 arcsecond clump of faint stars that do not seem to have any central condensation, which is a usual characteristic of a globular cluster (see Star Chart 2.24). The famous astronomer E. J. Hartung likened NGC 3201's stars to "short curved rays like jets from a fountain". Do you see this? Due to the fact that the globular lies near, and perhaps within, a large dust cloud, its brightness may be dimmed by as much as 2.2 magnitudes. Imagine what it would look like if the dust could be magically removed!

Our final object is the galaxy **NGC 3256**. This is only visible because the obscuring dust and gas of the Galaxy is thin enough to allow the distant galaxy's light to penetrate the Milky Way and reach us (see Figure 2.39). Located within a nice star field it appears as a bright oval-shaped object around 1.5 × 1 arcseconds in size. In a small telescope of 10 cm aperture it can be observed, although faintly (see Star Chart 2.25). Recent images indicate that it is undergoing some extreme tidal disruptions due to a galaxy merger, and has many streamers of material being drawn out of it.

The following constellations are also visible during these months at different times throughout the night. Remember that they may be low down and so diminished by the effects of the atmosphere. Also, you may have to observe them either earlier than midnight, or some considerable time after midnight, in order to view them.

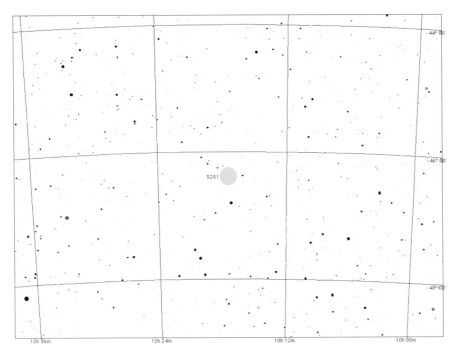

Star Chart 2.24 (*above*). NGC 3201.

Figure 2.38 (*right*). NGC 3201 (Space Telescope Science Institute, AAO, UK–PPARC, ROE, National Geographic Society, and California Institute of Technology).

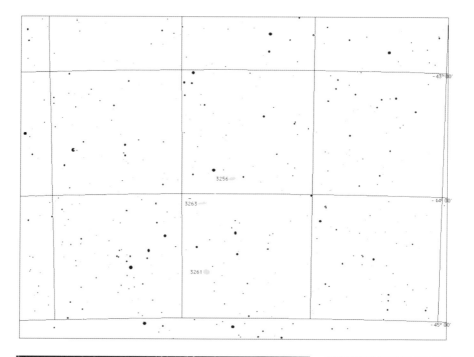

Star Chart 2.25 (*above*). NGC 3256.

Figure 2.39 (*left*). NGC 3256 (Space Telescope Science Institute, AAO, UK–PPARC, ROE, National Geographic Society, and California Institute of Technology).

Northern Hemisphere

Andromeda, Auriga, Camelopardalis, Canis Major, Canis Minor, Cassiopeia, Centaurus, Cepheus, Cygnus, Gemini, Hercules, Lacerta, Libra, Lupus, Lyra, Ophiuchus, Orion, Perseus, Scorpius, Scutum, Serpens Cauda, Taurus.

Southern Hemisphere

Apus, Ara, Auriga, Carina, Centaurus, Chamaeleon, Circinus, Corona Australis, Crux, Gemini, Libra, Lupus, Musca, Norma, Octans, Ophiuchus, Orion, Pavo, Perseus, Sagittarius, Scorpius, Scutum, Serpens Cauda, Telescopium, Triangulum Australe, Volans.

Objects in Monoceros

Stars

Designation	Alternate name	Vis. mag	RA	Dec.	Description
Epsilon (ε) Mon		4.5, 6.5, 5.6	$06^h 23.8^m$	+04° 36'	PA 27°; Sep. 13.4"[1]
15 (S) Mon		4.66$_v$, 7.5	$06^h 40.9^m$	+09° 54'	PA 213°; Sep. 2.8"
Beta (β) Mon		4.7, 5.2, 6.1	$06^h 28.8^m$	−07° 02'	PA 132°; Sep. 7.3"
Plasketts' Star		6.05	$06^h 37.4^m$	+06° 08'	Massive star

Deep-Sky Objects

Designation	Alternate name	Vis. mag	RA	Dec.	Description
NGC 2244	Herschel 2	4.8[3]	$06^h 32.4^m$	+04° 52'	Open cluster
NGC 2264	Christmas Tree	4.62	$06^h 41.1^m$	+09° 53'	Open cluster
NGC 2353	Herschel 34	7.1	$07^h 14.6^m$	−10° 18'	Open cluster
NGC 2506	Herschel 37	7.6	$08^h 00.2^m$	−10° 47'	Open cluster
NGC 2323	Messier 50	5.9	$07^h 03.2^m$	−08° 20'	Open cluster
NGC 2301		6.0	$06^h 51.8^m$	+00° 28'	Open cluster
IC 2177	Gum 2		$07^h 05.3^m$	−10° 38'	Emission nebula
NGC 2237–39	Rosette (Caldwell 49)		$06^h 23.3^m$	+05° 03'	Emission nebula
Cone Nebula			Just beyond tip of NGC 2264		Dark nebula
NGC 2261	Hubble's Variable Nebula (C 46)	10–13	$06^h 39.2^m$	+08° 244	Emission & reflection nebula

[1] Bear in mind that the position angle and separation may change over a short period of time, and so the values given here may be different from what you observe now.

[2] The subscript indicates the star is variable.

[3] The magnitude given for the deep-sky objects is the integrated magnitude, i.e. as if all the stars (or nebula) were concentrated to the size of one star.

Objects in Canis Minor

Stars

Designation	Alternate name	Vis. mag	RA	Dec.	Description
Alpha (α) Can Min	Procyon	0.40	07^h 39.3^m	+05° 13'	8th brightest star in sky

Deep-Sky Objects

Designation	Alternate name	Vis. mag	RA	Dec.	Description
NGC 2548	Messier 48	5.8	08^h 13.8^m	–05° 48'	Open cluster – missing Messier object?

Objects in Canis Major

Stars

Designation	Alternate name	Vis. mag	RA	Dec.	Description
Alpha (α) Can Maj	Sirius	–1.44$_v$	06^h 45.1^m	–16° 43'	Brightest star in sky
Sirius B		8.5	06^h 45.1^m	–16° 43'	PA 5°; Sep. 4.5"
Pi (π) Can Maj	Pup	4.7, 9.7	06^h 55.6^m	–20° 08'	PA 18°; Sep. 11.6"
Mu (μ) Can Maj		5.3, 8.6	06^h 56.1^m	–14° 03'	PA 340°; Sep. 3.0"
17 Can Maj		5.8, 9.3, 9.0	06^h 55.0^m	–20° 24'	PA 147°; Sep. 44.4"
Tau (τ) Can Maj		4.4, 10.5, 11.2	07^h 18.7^m	–24° 57'	PA 90°; Sep. 80.2"
UW Can Maj		4.7–5.0	07^h 18.7^m	–24° 34'	Variable star

Objects in Canis Majo (continued)

Deep-Sky Objects

Designation	Alternate name	Vis. mag	RA	Dec.	Description
NGC 2287	Messier 41	4.5	06^h 47.0^m	–20° 44'	Open cluster
Collinder 121		2.6	06^h 54.2^m	–24° 38'	Open cluster
NGC 2354		6.5	07^h 14.3^m	–25° 44'	Open cluster
NGC 2362	Caldwell 64	4.1	07^h 18.8^m	–24° 57'	Open cluster
NGC 2360	Caldwell 58	7.2	07^h 17.8^m	–15° 37'	Open cluster
IC 2165	PK 221–12.1	10.6	06^h 21.7^m	–12° 59'	Planetary nebula
NGC 2359	Duck nebula, Gum 4	–	07^h 18.6^m	–13° 12'	Emission nebula
IC 468		–	Northern end of NGC 2359		Emission nebula
NGC 2380		11.5	07^h 23.9^m	–27° 31'	Galaxy
Sh 2–308		–	06^h 54.2^m	–23° 56'	Bubble nebula around Wolf–Rayet star

Objects in Puppis

Stars

Designation	Alternate name	Vis. mag	RA	Dec.	Description
Δ 49	Dunlop 49	6.5, 7.2	07^h 28.9^m	–31° 51'	PA 53°; Sep. 8.9"
Sigma (σ) Pup	Dunlop 51	3.25, 8.6	07^h 29.2^m	–43° 18'	PA 74°; Sep. 22"
h 4046	Herschel 4046	6.2, 8.9	08^h 05.7^m	–33° 34'	PA 88°; Sep. 22"
h 4038	Herschel 4038	5.5, 8.5	08^h 02.7^m	–41° 19'	PA 346°; Sep. 27"
Beta (β) 454		6.4, 8.2	08^h 15.9^m	–30° 56'	PA 0.5°; Sep. 2"
H III 27	ADS 6255	4.5, 4.7	07^h 38.8^m	–26° 48'	PA 317°; Sep. 10"
L² Pup	HD 56096	2.6–6.2	07^h 13.5^m	–44° 39'	Variable star. M5IIIe

Deep-Sky Objects

Designation	Alternate name	Vis. mag	RA	Dec.	Description
Melotte 66		7.8	07^h 26.3^m	–47° 44'	Open cluster
NGC 2422	Messier 47	4.4	07^h 36.6^m	–14° 30'	Open cluster
NGC 2437	Messier 46	6.1	07^h 41.8^m	–14° 49'	Open cluster
NGC 2447	Messier 93	6.2	07^h 44.6^m	–23° 52'	Open cluster
NGC 2451		2.8	07^h 45.4^m	–37° 58'	Open cluster
NGC 2477	Caldwell 71	5.8	07^h 52.3^m	–38° 33'	Open cluster
NGC 2546		6.3	08^h 12.4^m	–37° 38'	Open cluster
NGC 2298		9.4	06^h 49.0^m	–36° 00'	Globular cluster
NGC 2438	PK 231+4.2	11.0	07^h 41.8^m	–14° 44'	Planetary nebula
NGC 2440	PK 234+2.1	9.4	07^h 41.9^m	–18° 13'	Planetary nebula
NGC 2452	PK 233–1.1	12.0	07^h 47.4^m	–27° 20'	Planetary nebula
NGC 2467	Gum 9	–	07^h 52.5^m	–26° 24'	Emission nebula
NGC 2525		11.6	08^h 05.6^m	–11° 26'	Galaxy

Objects Lepus and Columba

Stars

Designation	Alternate name	Vis. mag	RA	Dec.	Description
h 3857	Herschel 3587	5.6, 9.3	06^h 24.0^m	–36° 42'	PA 256°; Sep. 12.9"

Objects in Pyxis

Stars

Designation	Alternate name	Vis. mag	RA	Dec.	Description
I 489	ADS 6862	5.9, 6.5	$08^h 31.5^m$	–19° 35'	PA 338°; Sep. 0.24"
h 4166	Herschel 4166	6.7, 7.4	$09^h 03.3^m$	–33° 36'	PA 153°; Sep. 13.8"

Deep-Sky Objects

Designation	Alternate name	Vis. mag	RA	Dec.	Description
NGC 2613		10.4	$08^h 33.4^m$	–22° 58'	Galaxy
NGC 2888		12.3	$09^h 26.3^m$	–28° 02'	Galaxy
NGC 2818		8.2	$09^h 16.0^m$	–36° 37'	Open cluster
NGC 2818A	PK 261+8.1	11.6	$09^h 16.0^m$	–36° 38'	Planetary nebula

Objects in Antlia

Deep-Sky Objects

Designation	Alternate name	Vis. mag	RA	Dec.	Description
NGC 2997		9.4	$09^h 45.6^m$	–31° 11'	Galaxy

Objects in Vela

Stars

Designation	Alternate name	Vis. mag	RA	Dec.	Description
Gamma (γ) Vel	Dunlop 65	1.75ᵥ, 4.27	08ʰ 09.5ᵐ	−47° 20′	PA 220°; Sep. 41″
Delta (δ) Vel	Herschel 4136	1.99, 5.57	08ʰ 44.7ᵐ	−54° 43′	PA 154°; Sep. 2.2″
h 4104	Herschel 4104	5.48, 7.31	08ʰ 29.1ᵐ	−47° 56′	PA 244°; Sep. 3.4″
Δ 70	Dunlop 70	5.21, 7.09	08ʰ 29.5ᵐ	−44° 44′	PA 351°; Sep. 4.6″
Δ 81	Dunlop 81	5.76, 8.2	09ʰ 54.3ᵐ	−45° 17′	PA 239°; Sep. 5.3″
h 4245	Herschel 4245	6.8, 9.4	09ʰ 46.1ᵐ	−45° 55′	PA 216°; Sep. 9.4″
Rmk 13	Iᴬ and Iᴮ Vel	4.48, 8.38	10ʰ 20.9ᵐ	−56° 02′	PA 102°; Sep. 7.2″
Mu (μ) Vel	Russell 155	2.90, 5.92	10ʰ 46.6ᵐ	−49° 25′	PA 46°; Sep. 2.0″
Lambda (λ) 108		7.6, 9.6, 10.1, 9.2	08ʰ 57.1ᵐ	−43° 15′	Sep. AB 3.1″. Sep. AC 43.1″, Sep. AD 48.1″

Deep-Sky Objects

Designation	Alternate name	Vis. mag	RA	Dec.	Description
NGC 2547		4.7	08ʰ 10.7ᵐ	−49° 16′	Open cluster
IC 2391	Caldwell 85	2.5	08ʰ 40.2ᵐ	−53° 04′	Open cluster
IC 2488		7.4p[4]	09ʰ 27.6ᵐ	−56° 59′	Open cluster
NGC 3228		6.0	10ʰ 21.8ᵐ	−51° 43′	Open cluster
NGC 2660		8.8	08ʰ 42.2ᵐ	−47° 09′	Open cluster
Trumpler 10		5.1	08ʰ 47.8ᵐ	−42° 29′	Open cluster
IC 2395		4.6	08ʰ 41.1ᵐ	−48° 12′	Open cluster
Vela Supernova Remnant	Gum 16	–	08ʰ 30ᵐ	−44° 30′	Supernova remnant
NGC 2736.	The Pencil	–	09ʰ 00.4ᵐ	−45° 54′	Supernova remnant
RCW 40	Gum 25	–	09ʰ 02.4ᵐ	−48° 42′	Emission nebula
NGC 2792	PK 265+04.1	11.6	09ʰ 12.4ᵐ	−42° 26′	Planetary nebula

[4] This is the magnitude as estimated from photographic plates.

Objects in Vela (continued)

NGC 2899	PK 277–3.1	11.8	$09^h 27.0^m$	−56° 07'	Planetary nebula
NGC 3132	Eight-Burst Nebula	9.2	$10^h 07.7^m$	−40° 26'	Planetary nebula
NGC 2626		–	$08^h 35.6^m$	−40° 40'	Reflection nebula
NGC 3201	Caldwell 79	6.8	$10^h 17.6^m$	−46° 24'	Globular cluster
IC 2488	PK 285–14.1	10.4	$09^h 07.1^m$	−69° 57'	Planetary nebula
NGC 3256		11.3	$10^h 27.8^m$	−43° 54'	Galaxy

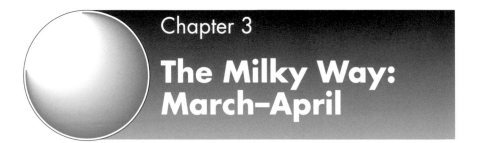

Chapter 3

The Milky Way: March–April

Carina, Crux, Musca, Centaurus, Circinus, Octans, Volans, Chamaeleon, Telescopium.
R.A 7^h to 15^h; Dec. –30° to –75°; Galactic longitude[1] 290° to 325°; Star Chart 3

3.1 Carina

We are now entering some very spectacular areas in the Milky Way that contains equally spectacular objects. It is a cruel twist of fate that observers in the northern hemisphere are unable to see these objects, so my advice to these observers would be either to plan an observing trip to the southern skies, or do the next best thing and stay at home and read this book.

Our first constellation is **Carina**, a large and important one (see Star Chart 3.1). It can be likened to Cassiopeia in the northern skies as it is circumpolar for southern observers and thus never sets below their horizon. The transit of Carina is something of a problem to this author because it has such a large range in RA, so I have decided to put it in this section, as opposed to the previous section, as much of what we will observe is best placed toward the end of February and early March. The Milky Way is very rich here, and the constellation is full of clusters, emission and reflection nebulae, planetary nebulae and multiple stars. In fact, a considerable portion of this book could be spent on this area alone, to the detriment of other constellations. Some 50 star clusters are found here, and I shall just discuss those brighter and more famous ones. It is extremely rewarding to scan this part of the sky with binoculars, as there will always be something to see and to take your breath away. Many hours can be fruitfully spent just lying back and looking.

Let's begin with some nice double stars. The system **h (Herschel) 4130** is a nice double star of magnitudes 6.5 and 8.4. They present a nice color contrast of a very pale yellow (or even white) and red. The backdrop is a lovely star-strewn field. A fine object for small telescopes is **Upsilon (υ) Carinae**, with magnitudes 3.08 and 6.25, and near this pair at around 5 arcminutes to the southeast is an even closer double, **h (Herschel) 4252**, also easily seen. A little-known star and probably missed off many an observing schedule is **Hrg (Hargreave) 47**. This is one of those systems that little is known about, as not a lot has

---

[1] See Appendix 1 for details on astronomical coordinate systems.

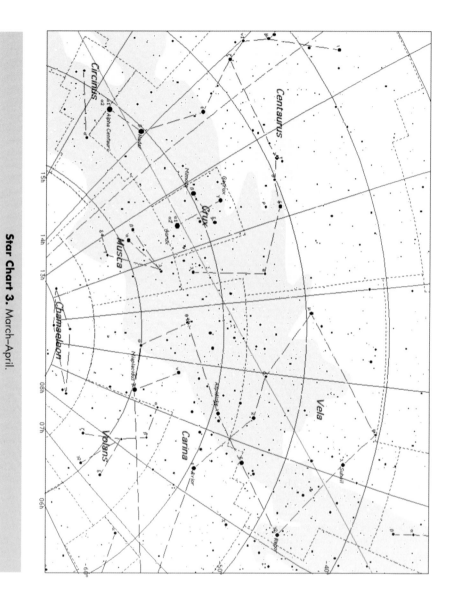

Star Chart 3. March–April.

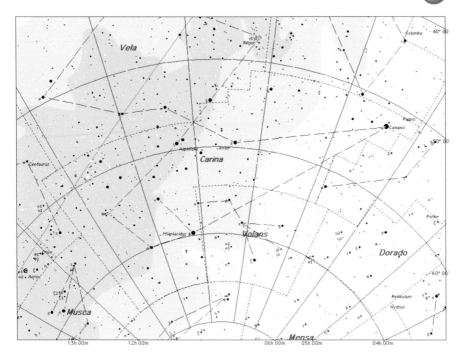

Star Chart 3.1. Carina.

changed over the years, with the separation and position angle remaining more or less constant. Nevertheless, with a 15 cm telescope it does provide a rather nice view. Set in a spectacular star field is t^2 **Carinae**. Located just northwest of the famous **Eta (η) Carinae** (see entry later), this is a bright orangish star with a fainter white secondary. Even a small telescope will spit the system, but what is truly impressive is the backdrop of arcs and groups of stars. Another fine double star is **Gli 152**. Though unequal in brightness, they make a fine view consisting of a pale orange and a white star. The emission nebula **NGC 3324** is a short distance away some 15 arcminutes to the northwest. A fitting finale to our list of double stars is the close system **R 164**, a nice bright yellow primary with a fainter white secondary. Even a small telescope of, say, 10 cm will split the system, although steady seeing conditions will be needed.

There are a few variable stars in this area of the Milky Way, and we shall look at just two. The giant star **R Carinae** is a long-period variable of Mira type. It is, as expected, a bright orange-red color, and when it reaches maximum, it truly stands out amongst the nice sprinkling of background stars. It has a period of around 309 days. **S Carinae** is also a Mira-type long-period variable, again orange-red in color. It too is set amongst a nice backdrop of stars.

Nebulae of all sorts abound in Carina: emission, reflection, and planetary, and the constellation makes an ideal target for astrophotographers and CCD imagers. But there is also plenty for the visual observer.

The reflection nebula **IC 2220** is a famous object and often called the **Toby Jug**, or **Butterfly Nebula**, due to its bipolar shape. A telescope with an aperture of 20 cm will show the nebula as a faint haze surrounding the irregular red variable star **V341** especially to its southwest and northwest (see Star Chart 3.2). Using a larger aperture the famous shape becomes apparent (see Figure 3.1). Incidentally, the star varies in magnitude from 6.1 to 7.06.

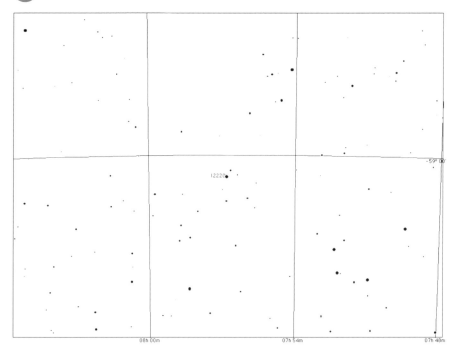

Star Chart 3.2. IC 2220.

Figure 3.1. IC 2220 (Space Telescope Science Institute, AAO, UK–PPARC, ROE, National Geographic Society, and California Institute of Technology).

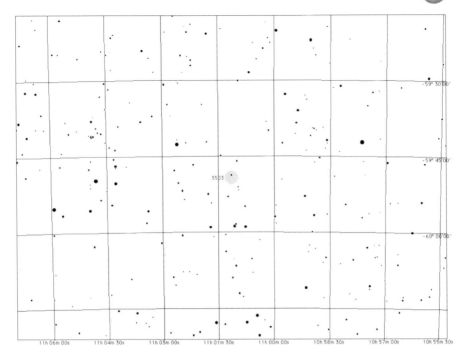

Star Chart 3.3. NGC 3503.

Another reflection nebula is **NGC 3503**, located within a rich star field and around 1.5 arcminutes across. A telescope of 15 cm aperture should have no difficulty with this object (see Star Chart 3.3). As a bonus, there are four faint 10th and 11th magnitude stars seemingly within the nebula forming a small asterism.

There are so many emission nebulae in Carina that it is difficult to know what to mention and what to leave out. The following examples are only a representative selection.

First discovered in 1834 by John Herschel, **NGC 3199** is a large emission nebula. It appears as a broad crescent-shaped object, some 7 × 3 arcminutes in size, with its north-eastern section particularly well defined, as it has a dark area that makes the rest stand out in comparison. There are many double and triple stars in the field of view that all in all go to make a very nice object. A 15 cm telescope will show the basic shape of the nebula, but using the [OIII] filter will allow smaller telescopes to enjoy the view (see Star Chart 3.4).

Another object discovered by John Herschel is **NGC 3247**. He described it as a "decid-edly nebulous patch". However, the NGC reference seems to be in error. If the coordinates given in the catalogue are used, one finds a small cluster of about 25 stars instead. The object Herschel referred to is some 15 arcminutes to the northwest. Using a 15 cm tele-scope, the nebula is easily observed, some 5 arcminutes in size. The use of an [OIII] filter will greatly enhance the view (see Star Chart 3.4).

A telescope of aperture 20 cm or more, with a high magnification, shows a compact and small cluster at the nebula's eastern side, separated from it by a thin band of dark material. The cluster is known as **Westerlund 2**, and it is believed that its hot and luminous stars are responsible for ionizing the nebula's gas. Dark dust clouds are particularly thick here, and extinguish the light from the nebula by about 5 magnitudes. Just imagine the glorious view we would have if the dust were absent!

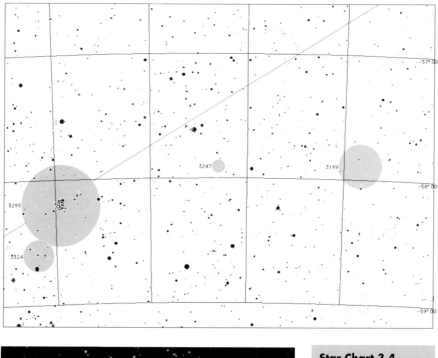

Star Chart 3.4
(*above*). NGC 3199,
NGC 3247, NGC
3293, NGC 3324.

Figure 3.2 (*left*).
NGC 3324 (Space
Telescope Science
Institute, AAO,
UK–PPARC, ROE,
National Geographic
Society, and California
Institute of Technology).

A much larger nebula is **NGC 3324** (see Figure 3.2). It is fairly spread out, and so has no area of particular concentration, although its northwestern edge seems brighter. It may be, and research indicates this may well be the case, that it is in fact an outlying part of the vast **Eta Carinae** nebula complex (see Star Chart 3.4). Lying as it does in a nice star field, it also contains the pair of stars **h** (**Herschel**) **4338** at magnitudes 8.2 and 9.5, respectively. Both stars are of type O, and are responsible for the ionization of the nebula.

A favorite on many observers' list are the nebulae **NGC 3581–82**. Yet another discovery by John Herschel in 1834, they are the brighter members of a group of nebulous objects located in a rich star field (see Figure 3.3). They form a small fan or wedge shaped object, some 7 × 5 arcminutes, and will appear as a hazy wisp of milky light. It is worthwhile looking carefully around the field of view, as there are several other fainter nebulae here. They may all be regarded as part of one larger nebula that is crossed by several dark dust lanes, thus giving the impression of many separate nebulae (see Star Chart 3.5). The northern section is cataloged as **NGC 3579** and **NGC 3584**. The two detached and quite faint nebulae lying to the southwest and southeast are **NGC 3576** and **NGC 3586**, respectively. It is a very worthwhile object to observe, but be warned – you will need a medium-sized aperture telescope and it goes without saying that a filter will greatly help.

A mysterious object that is located within the same field of view as NGC 3581 is the emission nebula **NGC 3603**. This will appear as a small 2 arcminute hazy object. What is especially interesting is that it is in fact part of a vast HII region that is located in the **Carina Spiral Arm** of our Galaxy. It is, however, heavily obscured by the intervening gas and dust, and lies about 8000 parsecs away. With a large telescope, a faint central star can be glimpsed that is in reality the very condensed core of the object. A small telescope will locate the nebula, but for a detailed view, large apertures are essential.

We now turn our attention to one of the most spectacular objects in the entire sky, the **Eta (η) Carinae Nebula** complex (**Caldwell 92**). The star itself is a most unusual variable

Star Chart 3.5. NGC 3581.

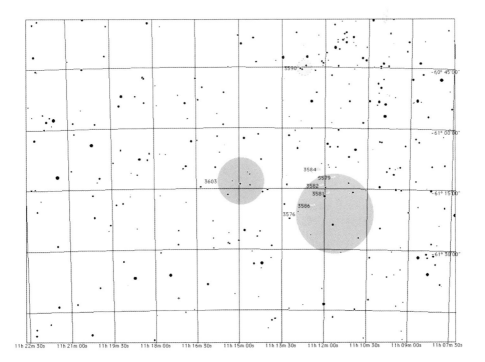

Figure 3.3. NGC 3581 (Space Telescope Science Institute, AAO, UK–PPARC, ROE, National Geographic Society, and California Institute of Technology).

star that has a very odd[2] observational history. It all starts in 1677, when Edmund Halley first classified it as 4th magnitude. It then varied erratically in brightness over the next century, from 2nd magnitude in 1720, to 4th magnitude again in 1782. It repeated this cycle up until 1820. The star then began to brighten rapidly and was 2nd magnitude by 1822 and 1st magnitude in 1827. By the time 1827 arrived it was as bright as **Canopus**.[3] However, by 1868, it had faded completely out of view. At the beginning of the 20th century, it reappeared as an 8th magnitude star, and now varies between 6th and 7th magnitude. So, what is happening here? It seems that the star, while not in its death throes, is still getting on a bit, and going through the latter stages of its stellar evolution.[4] Many astronomers predict that it will soon (or maybe not) explode as a supernova. If that happens, it will be a fantastic sight, and once again southern astronomers will have a glorious time observing another supernova!

Telescopically, the star is nothing special, appearing as an orange blob. However surrounding it is a most spectacular nebula, one of the finest in the entire sky (see Figure 3.4). The nebula, also called the **Keyhole Nebula,** is one of the brightest and largest of its kind. It is made up of four large patches of nebulosity that are divided by wide and dark dust lanes. In fact it is so bright that each section can be glimpsed with the naked eye. It comes as a surprise to many amateur astronomers when they learn that the nebula is far brighter than the more famous Orion Nebula, and that it covers a vast 120 arcminutes of sky. Even on the most ordinary of nights, it can be seen as a faint hazy patch of light, and binoculars will show several triangle or wedge-shaped patches of nebula crossed by very dark dust lanes. The nebulosity and dust lanes extend far beyond what can be glimpsed in the field of view,

---

[2] An understatement!

[3] Canopus is the second-brightest star in the sky and is a glorious sight. However, as it does not lie in any part of the Milky Way, we do not discuss it further, alas.

[4] A good book on stellar evolution that is geared towards the amateur astronomer is *An Observer's Guide to Stellar Evolution* (Springer-Verlag) by the present author.

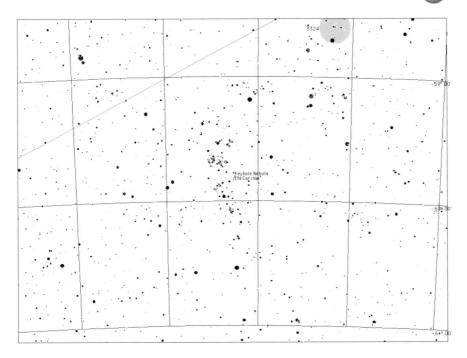

Star Chart 3.6. Eta Carinae and surrounding nebulae and clusters.

and deep imaging of the region shows it covers more than 4 square degrees (see Star Chart 3.6).

The bright orange star Eta Carinae is surrounded by an orange nebula about 15 arcseconds in size called the **Homunculus Nebula**. A small telescope with medium magnification will show this, but larger telescopes will resolve much more of its structure, and under good conditions the bright star-like center, some 2 arcseconds across, may be seen. Such is the brightness of the Homunculus, that it may be a very good object for those amateurs who express an interest in obtaining the spectrum of such objects.

Several star clusters reside in or near the nebula. The small group that surrounds Eta Carinae is called **Trumpler 16**, while the compact group some 10 arcminutes to the northwest is called **Trumpler 14**. At the center of this latter cluster is the nice double star **h 4356**, magnitudes 7.2 and 8.9, separated by the close distance of 2.9 arcseconds. You should be able to see these stars in an 8 cm telescope. As an aside, it is interesting to note that the brighter of the two stars is **HD 93129A**, and is believed to be one of the most massive and luminous stars known. Then, located about 2 arcminutes southeast is another double star system, **h (Herschel) 4360**, both of magnitude 7.7 and separated by a wide 12 arcseconds. If the seeing conditions allow, you may also notice that the northwest star is itself a close double.

You will not be surprised to know that the distance estimates to the Eta Carinae Nebula vary greatly from 2000 to 3200 parsecs. Take your pick.

Many planetary nebulae reside within Carina. Most are faint, but there are one or two bright examples. Our first such example is **IC 2448**, at about 10 arcminutes across, slightly oval in shape and located within a field of stars. A small telescope of, say, 8 cm will show it, but an [OIII] filter really does help here. A nice pale greenish-blue planetary nebula is **NGC**

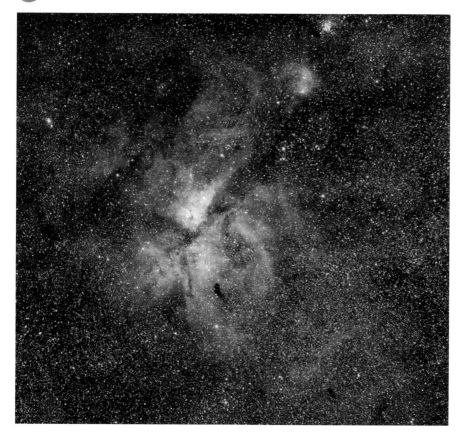

Figure 3.4. Eta Carinae (Matt BenDaniel, http://starmatt.com).

2867 (**Caldwell 90**). In a small telescope it will appear as a slightly out-of-focus star, and with medium aperture and magnification is easy to find. It is only 12 arcseconds across, and the central star, which incidentally, is a very hot (80,000 K) Wolf–Rayet type, is a very faint 14th magnitude, so is a very difficult object to find visually. Some planetary nebulae will appear star-like, and such an object is **IC 2501**. Providing you use a high enough magnification, then the disk, of only 5 arcseconds diameter, will be resolved. Another star-like planetary nebula is **IC 2553**. Under high magnification its 4 arcsecond disk will be immediately obvious against the rich background of stars. **NGC 3211** is a lovely pale blue, about 12 arcseconds in diameter with no visible central star (see Star Chart 3.7). It can be resolved with a 10 cm telescope, but smaller apertures will need very good seeing.

The planetary **IC 2621** is a difficult object to resolve and may appear stellar in all but the largest apertures. With excellent conditions and at least a 20 cm aperture, a minute 5 arc-second bluish disk may be glimpsed.

Something of a conundrum is the elliptical ring nebula surrounding the star **AG Carinae**. Discovered in 1950, it is fairly large, at 40 × 30 arcseconds. The nebula seems to have been ejected by the star AG Carinae, which is a very luminous P Cygni-type supergiant that varies from 6.17 to 8.5 magnitude. It seems very similar to Eta Carinae and its attendant Homunculus. Is it a planetary nebula or the precursor to one, or just a nebula

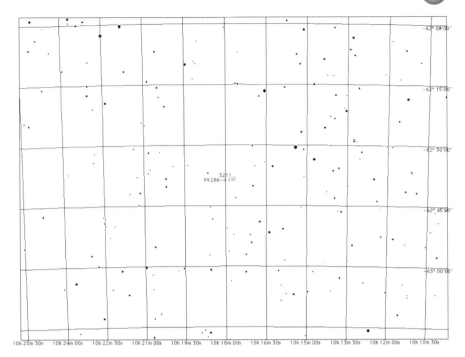

Star Chart 3.7. NGC 3211.

surrounding a star? Whatever its origins, the nebula is a difficult object to observe due to the glare from the star, and a 30 cm telescope will show it providing the conditions are optimal. Some amateurs suggest the use of an Hβ filter and observe when the star is at minimum light.

Besides being full of nebulae and wonderful stars, Carina also has a lot to offer in the way of star clusters. One of the nicest is **NGC 2516 (Caldwell 96)**, which has about 80 stars of magnitude 7. At a diameter of 50 arcminutes it is nearly the size of the full moon, and appears to the naked eye with an integrated magnitude of 3.8 (see Figure 3.5).

It is best seen under low magnification to get the full impact, and there are three bright pale-orange stars that contrast nicely with the other blue and white cluster members. The cluster can be easily located as it lies about 3° to the southwest of **Epsilon (ε) Carinae**, and is always a favorite with astrophotographers.

One cluster that is perhaps best suited to smaller telescopes is **NGC 3114**. It has about 50 stars spread over a region 35 arcseconds in diameter. The total magnitude of the cluster is 4.2, and it is interesting to compare this cluster with the previous one, NGC 2516, to see which one you find the most pleasing. The cluster **NGC 3293**, also called the **Gem Cluster**, is a very nice group to look for (see Figure 3.6). It is bright but small, at about 6 arcseconds in diameter, and has a bright red supergiant star at its center which makes a nice contrast with the other members

A famous cluster is **IC 2602 (Caldwell 102)**, which has been called the **Southern Pleiades**, and the **Theta Carinae Cluster**. It is perfect for binoculars and small telescopes. It is over 1° in diameter and has about 60 stars in it. Residing in the cluster is the star **Theta (θ) Carinae** which shines at magnitude 2.74, and this makes the total magnitude for the cluster 1.9. This means that it is easily visible to the naked eye (see Figure 3.7).

Figure 3.5. NGC 2516 (Space Telescope Science Institute, AAO, UK–PPARC, ROE, National Geographic Society, and California Institute of Technology).

Our last open star cluster is the wonderful **NGC 3532** (**Caldwell 91**), which John Herschel once described as "the most brilliant object of its kind I have ever seen". While you may not agree with his words, it is without a doubt a spectacular object. It is a very large cluster, some 55 arcseconds in diameter, and has a total magnitude of 3, so it is visible to the naked

Figure 3.6. NGC 3293 (AAO, UK–PPARC, ROE).

Figure 3.7. IC 2602 (IAS).

eye even on those nights when conditions are not at their best for deep-sky observing. It contains many doubles and triples and stars that are seemingly not placed at random, but in arcs and streams. All the stars are uniformly white in color, except a pale-orange star, the 6th magnitude **HD 98544**, which is also the brightest cluster member. All in all this is a very nice cluster (see Figure 3.8).

Our final object in Carina is the globular cluster **NGC 2808**. Set against a lovely backdrop of stars, this is a perfect example of a type of globular that has a symmetrical structure and is highly compressed (see Figure 3.9). With a diameter of around 5 arcseconds, a 15 cm telescope should be able to resolve it. In smaller telescopes and even binoculars, although the stars will not be resolved, it will still show as a bright fuzzy spot with a strong concentration towards the center.

3.2 Crux

The smallest constellation in the sky, **Crux**, only covers an area of about 68 square degrees, yet contains some of the most famous objects in the entire sky (see Star Chart 3.8). Its midnight culmination is about 30 March, and thus it is an autumn constellation for southern observers. It contains not only one of the most colorful star clusters known, but also an equally famous dark nebula. Not only that, it has such a distinctive constellation shape that the **Southern Cross** is taken by many to symbolize the southern hemisphere.

The Milky Way completely engulfs Crux, and the whole region is perfect for sweeping with binoculars (see Figure 3.10). Alas, it is too far south for northern observers, and only those regions in the USA at around latitude +30° can ever hope to glimpse this wonderful part of the Milky Way.

Let's start by looking at some stars. **Alpha (α) Crucis** is a bright double star of magnitudes 1.25 and 1.64. The colors are white and blue-white respectively. There is also a third star some 90 arcseconds distant, at a magnitude of 4.86. All three can be easily seen in an

Figure 3.8. NGC 3532 (IAS).

8 cm telescope, but as the brighter pair are only 4.4 arcseconds apart they are too close to be resolved with binoculars. Evidence points to the stars being part of a true physical system, although there has been little change in their separation for nearly 200 years.[5] Therefore they must all have very long periods. Alpha Crucis lies in a very rich star field, and there are some who say it can be glimpsed during daylight hours. What can you see?[6]

Another star in the Southern Cross is **Gamma (γ) Crucis**, which is a wide optical double star with a bright orange primary and white secondary star, magnitudes 1.59 and 6.42, respectively. The companion star lies some 2 arcminutes away. The third star in the cross that we shall discuss is **Beta (β) Crucis**, which is a multiple star. It has a 1.25 magnitude primary, and, some 44 arcseconds away, a faint 11th magnitude white star that is just a field star. What makes observing this otherwise ordinary double special is the lovely 7th magnitude deep red carbon star **EsB 365**. This contrasts wonderfully with the blue-white primary. These two stars can be seen in small telescopes, but to see the much fainter star a 25 cm aperture will be needed. Our final double is the nice **Mu (μ) Crucis**. This easily split wide pair of stars, magnitudes 4.03 and 5.08, are separated by 35 arcseconds and so can easily be resolved in almost any telescope.

---

[5] Recent work has suggested that the separation of the bright pair is very slowly decreasing.
[6] Be careful when observing during the day. Watch out for that big bright yellow thing we astronomers call the Sun! I say this because I am constantly amazed by how many amateur astronomers fail to take adequate and sensible precautions whilst observing during the day. **Never point any optical equipment at the Sun, and never under any circumstances look at the Sun without proper and safe solar protection.**

Figure 3.9. NGC 2808 (Space Telescope Science Institute, AAO, UK–PPARC, ROE, National Geographic Society, and California Institute of Technology).

Star Chart 3.8. Crux.

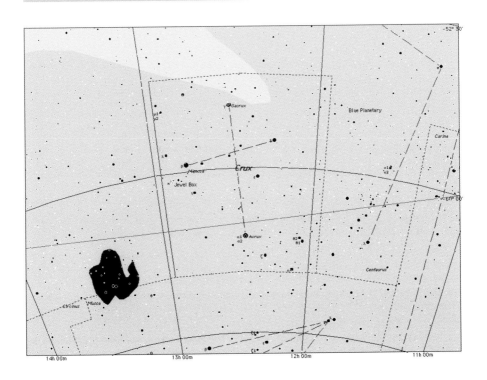

Figure 3.10. Crux; Coalsack (Matt BenDaniel, http://starmatt.com).

There is also a Cepheid variable star in Crux that can be observed and this is the star **R Crucis**. It varies in magnitude from 6.4 to 7.2 with a period slightly less than 6 days. It is a perfect variable star for observation with binoculars or a small telescope.

Let's now look at that object for which Crux is justifiably famous, the **Jewel Box Cluster**. The cluster, **NGC 4755 (Caldwell 94)**, gets its name from John Herschel's description "a casket of variously colored precious stones", and indeed does show a selection of bright and delicate colors (see Figure 3.11). It is also known as the **Kappa Crucis Cluster** after the brightest star it contains. Many amateurs who have had the lucky opportunity to observe clusters in both

Figure 3.11. NGC 4755 (IAS).

hemispheres often say that this one is the finest. However, it is a surprisingly small cluster, only 10 arcminutes across. The colors are not vivid except for one bright orange-red super-giant star **Kappa (κ) Crucis**, but more of a pale and delicate hue, but color they defiantly do exhibit. However, the use of larger apertures will enhance the colors of the stars.

The stars are in a marked geometric pattern with its three brightest members in a line stretching from the northeast to the southwest. These three stars have often been referred to as the traffic lights due to their contrasting colors. The cluster stars range in magnitude from 6th to 10th. It is just visible to the naked eye and in binoculars the cluster is great; in large apertures it is wonderful. The estimated distance to the Jewel Box is about 2400 parsecs, but this may be in considerable error as its light may be heavily obscured by an object we will discuss next – the Coalsack.

Figure 3.12. NGC 4103 (Space Telescope Science Institute, AAO, UK–PPARC, ROE, National Geographic Society, and California Institute of Technology).

The Milky Way is very bright and conspicuous in Crux, and so any dark dust cloud will stand out in comparison. Such a dark cloud is the **Coalsack** (**Caldwell 99**). This dark nebula, lying to the east of Alpha Crucis, is some 7° × 4° in size and lies only 180 parsecs away. It probably shares, along with the Great Rift in Cygnus, the distinction of being the most famous of all dark nebulae. Set against the Milky Way, it really does look like a big hole and can easily fill the entire field of view of small binoculars. At first glance it will seem to be an amorphous and plain feature, but under superb seeing conditions and with clean optics, a delicate structure can be glimpsed within the darkness. Also, one fallacy is that there are no stars visible in the Coalsack. This is just not true. It appears totally black to the naked eye due to the incredible brightness of the surrounding Milky Way in which it is situated. Check for yourself. It may well be that the Coalsack is the nearest cloud of obscuring nebulosity to the Earth (see Figure 3.10).

There are several bright open star clusters in the constellation. **NGC 4103** is a nice group of about 50 stars that can easily be viewed with binoculars or rich-field telescopes (see Figure 3.12). It is around 2 arcminutes across and contains several 9th magnitude stars set against the hazy remainder of stars.

Situated mid-way between Alpha Crucis and **Epsilon (ε) Crucis** is the beautiful cluster **NGC 4349**. These bright stars set in a group 20 arcminutes in diameter, with an integrated magnitude of around 7.4, are visible as a misty glow in small telescopes (see Figure 3.13). Individual stars are around 11th magnitude, and so will not be resolved with binoculars.

Several star clusters are often overlooked by amateurs as they are faint, but this shouldn't be a deterrent. Amongst them is **NGC 4052**, located northwest of the double star **Theta (θ) Crucis**. Although faint, it is a nice grouping of around 75 stars. A cluster that stands out from the dark Coalsack is **NGC 4609** (**Caldwell 98**). It is small and has around five stars of magnitude 9 (see Star Chart 3.9). Binoculars do not show much, and even an 8 cm telescope will only show about 20 stars. What is really difficult to see is the tiny cluster **Hogg 15**, which is about 12 arcminutes southeast of NGC 4609. It is only 2 arc-minutes across and so will need at least a 30 cm telescope to be fully appreciated or even found! Passed over by many is **Ruprecht 98**. This tiny cluster will appear as a 7th

Figure 3.13. NGC 4349 (Space Telescope Science Institute, AAO, UK–PPARC, ROE, National Geographic Society, and California Institute of Technology).

Star Chart 3.9. NGC 4609.

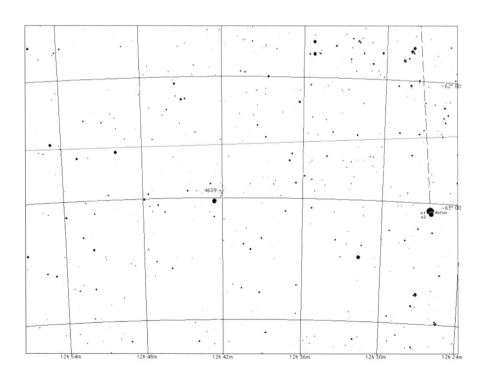

magnitude hazy glow buried amongst the vivid star fields of the Milky Way. Small binoculars will just find it, while larger ones will resolve its 60 members. It can be found in the far southwestern corner of the constellation, about 1° west of **Eta (η) Crucis.**

3.3 Musca

The small constellation **Musca** lies directly south of Crux, and is circumpolar for many, if not all, southern hemisphere amateur astronomers (see Star Chart 3.10). It too lies within the Milky Way. Obscuration by dust and gas is believed to be heavy here as there are no galaxies visible, at least to amateur instruments, and as we shall see, even the star clusters are small and faint. But there are plenty of objects for us to observe.

Many double stars are to be found here and we will look at just a few. The first is **h (Herschel) 4432**, which is a nice pair of yellow-white stars at 5.36 and 6.68 magnitudes, separated by 2.3 arcseconds. The actual colors are in doubt as some amateurs do not see any yellow, but rather all white. A nice double is **Beta (β) Muscae.** This white double star system is separated by 1.3 arcseconds and so is a difficult pair to resolve. The magnitudes are 3.51 and 4.01. The pair is opening and so should make splitting easier, but only time will tell. A famous double star for several reasons is **Theta (θ) Muscae.** The stars are a physical pair and shine at magnitudes 5.64 and 7.66, separated by about 5.5 arcseconds. What is so special is that the brighter of the two is a spectroscopic binary star and also a Wolf–Rayet binary system, the second-brightest in the sky after γ^2 Velorum (see earlier

Star Chart 3.10. Musca.

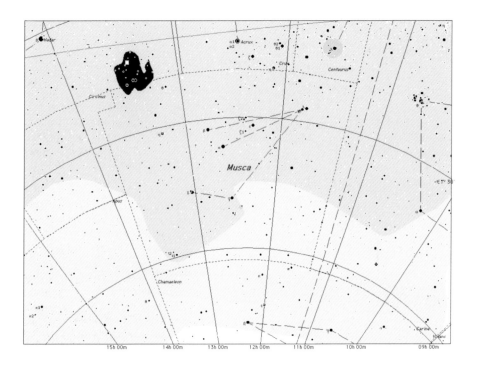

Figure 3.14. NGC 4463 (Space Telescope Science Institute, AAO, UK–PPARC, ROE, National Geographic Society, and California Institute of Technology).

entry in Vela). It is classified as a Wolf–Rayet star, indicating that it is one of the rare carbon stars. Some observers are quoted as saying that it is a fine red-orange color that contrasts nicely with its white companion B0-type star. Others report seeing no color contrast. Something of a strange situation. What do you see?

Contrary to popular belief, there are some open star clusters that can be seen with amateur equipment. One of these is **NGC 4463**, found on the southwestern edge of the Coalsack, and thus, although it is very small, it can be found relatively easily due to the contrast with the surrounding blackness (see Figure 3.14).

With small binoculars it will appear as a faint blob, and in larger binoculars or small telescopes will be resolved into only two stars surrounded by a faint milky haze (see Star Chart 3.11). Another very faint cluster is **NGC 4815**, again located in the Coalsack. However, this time large-aperture telescopes will be needed to resolve the cluster's stars, and in binoculars and small telescopes it will remain a very small hazy spot.

There are also some globular clusters that can be seen. One of these is **NGC 4372**, found just southwest of **Gamma (γ) Muscae**. It is partially obscured by dust lanes, but still appears as a large object some 10 arcseconds in diameter. In binoculars it will appear as a faint disk but in larger instruments a few faint stars can be seen. Resembling a comet is the globular cluster **NGC 4833**. This is a nice large, but fairly faint object lying just less than 1° from **Delta (δ) Muscae**. In binoculars it really does look like a small faint comet, but with telescopes of aperture 20 cm and larger, its true nature becomes apparent (see Figure 3.15). It is around 4 arcminutes in diameter at a distance of 5500 parsecs (see Star Chart 3.12 for both NGC 4372 and NGC 4833).

Several nebulae also make an appearance in Musca, and none more so than **NGC 5189**, which is surrounded by controversy. It is bright and easily found, located some 5° north-east of Beta Muscae (see Star Chart 3.13). John Herschel called it "a very strange object". It lies within a lovely star field and is about 1.5 by 1 arcminutes in size, with an irregular structure. Some astronomers believe it is an emission nebula, others a planetary nebulae. Recent indications favor the latter conclusion. Due to its slight resemblance to a spiral

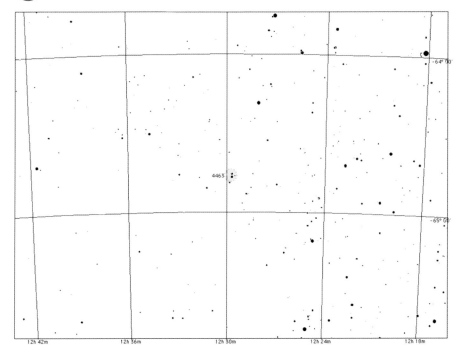

Star Chart 3.11. NGC 4463.

Figure 3.15. NGC 4833 (Space Telescope Science Institute, AAO, UK–PPARC, ROE, National Geographic Society, and California Institute of Technology).

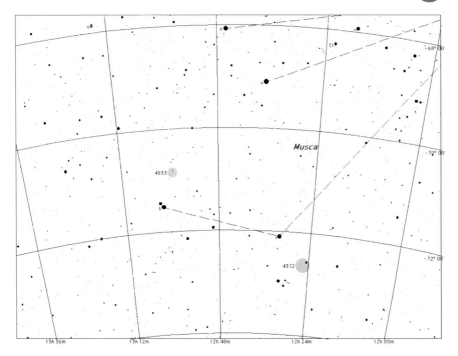

Star Chart 3.12 (*above*). NGC 4372; NGC 4833.
Star Chart 3.13 (*below*). NGC 5189.

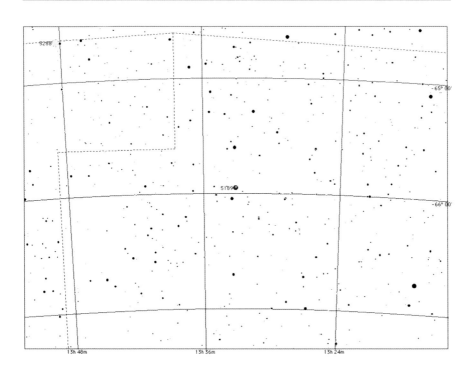

Figure 3.16. NGC 5189 (Space Telescope Science Institute, AAO, UK–PPARC, ROE, National Geographic Society, and California Institute of Technology).

galaxy, it has been given the name the **Spiral Planetary** (see Figure 3.16). An [OIII] filter greatly helps in observing this object.

Three other planetary nebulae can also be found in Musca. **NGC 4071** is a difficult object to observe, even in a 30 cm telescope, but with the addition of an [OIII] filter, it becomes much easier. At about 1 arcminute across, it has an even surface brightness, although some amateurs report that the preceding edge is brighter. **IC 4191** is a very small planetary nebula only 5 arcseconds in diameter. It appears as a tiny pale blue disk set against a lovely star field. It may be wise to use the 6th magnitude orange star **HD 113919** as a guide in locating this object, which is 9 arcminutes south of the nebula. In a small telescope it will be decidedly stellar in appearance.

Our final object in Musca is the planetary nebula **MyCn 18**. This is an exceedingly difficult object to find and observe, and it can be said that, in truth, you will only locate it if you use an [OIII] filter. At magnitude 12.5, set in a star-strewn field, it is nigh on impossible to find visually without the filter; otherwise it will just appear star-like. A faint, wide double star about 1.5 arcseconds northwest can be used as a guide. A definite challenge I think!

3.4 Centaurus

One of the southern skies' premier constellations is **Centaurus**. The Milky Way is dense here and only the northwestern part of the constellation can be considered "outside" of it. The northern boundary of Centaurus is at latitude –30° which means that it is visible from well up into the northern hemisphere (see Star Chart 3.14). However, the southern boundary reaches to –65°, so can only be seen by those lucky enough to live south of +20° (see Figure 3.17). It transits during early April.

Centaurus is a constellation of superlatives: the best globular cluster, the nearest star to the Solar System, the third-brightest star in the sky. It also contains, as we have come to

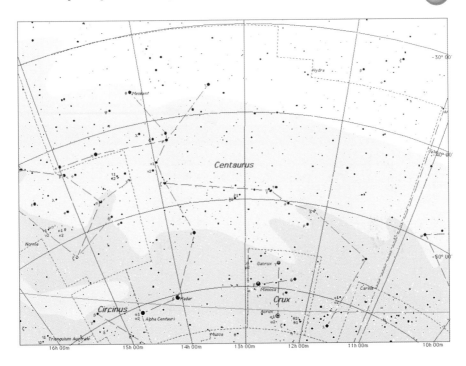

Star Chart 3.14. Centaurus.

expect from a Milky Way constellation, plenty of double stars, nebulae, open and globular clusters, and considering we are deep in the Milky Way, several galaxies, including one that is very peculiar! So without much ado, let's start.

Double stars abound here and we will look at just the brighter ones. A nice pair of yellowish stars is **h (Herschel) 4423**. Its separation and position angle are slowly increasing and it is an easy object for an 8 cm telescope. An easy orange binary system is **Brs 5**, which is set in a nice star field. It too has a separation that is slowly increasing and it can be split in a small telescope. A nice color contrast can be seen in **D Centauri (Rmk 14)**, of pale orange and white. The magnitudes are 5.75 and 6.98, respectively. As to be expected, considering it is a rich area of the Milky Way, it has a backdrop of field stars. A difficult binary to split is **Gamma (γ) Centauri**. A 15 cm telescope will be needed to split this, and perhaps even larger as the separation was only 1 arcsecond in 2000 and never gets above 1.8 arcseconds. Both stars are magnitude 2.9, and a lovely yellow color. Another double that may prove difficult to resolve is **I 424**, which consists of a pale yellow star (some would say white), and a white companion very close to the north. It can just be seen in a 10 cm telescope. The brightest star is an extremely close binary in itself, **See 170**, and will need a very large aperture to split, if it can be split at all. About 8 arcminutes to the northeast is a wide pair of differing magnitudes colored orange and pale red, **CorO 152**.

A lovely white system placed in a star-strewn field is **Δ (Dunlop) 141**. First discovered by John Herschel in 1835, it is a fine object for small telescopes with magnitudes 5.24 and 6.63. A fairly obscure double star is **R (Russell) 223**. It consists of a nice orange star set in a very profuse star field and its companion is a faint white star. A small telescope will just resolve this system. It is one of those many double-star systems we come across from time to time

Figure 3.17. Centaurus to Carina (Matt BenDaniel, http://starmatt.com).

that has yet to be studied in any detail. On the other hand, the double star **CorO 157** has been studied in detail. It is an orange star with a faint tinted companion whose color has been described as "ashy". What do you think it is? The secondary is well separated to the northwest, and is a difficult object in, say, a 6 cm telescope, but easy in 10 cm and larger. The brighter star is apparently a hypergiant that shows some slight variability, while the companion is a B-type star that is credited to be exciting the very faint emission nebula **RCW 80**.

A fine color-contrasted pair is **Howe 24**. The stars are deep yellow and red. They are well separated but the separation is decreasing slowly. A lovely pair of white stars with magnitudes 4.51 and 6.45 is **3 Centauri (Δ 148)**. Our penultimate double star is **Beta (β) Centauri**. This is a lovely bright blue star that was once regarded as a single. It is, however, a very difficult double star to split. The eleventh-brightest star in the sky, and unknown to northern observers because of its low latitude, lying as it does only 4.5° from Alpha Centauri (see below), it has a luminosity that is an astonishing 10,000 times that of the Sun. A definitely blue-white star, it has a companion of magnitude 4.1, but is a difficult double to split as the companion is only 1.28 arcseconds from the primary. A 20 cm telescope will be needed in order to resolve the system. Deep images of the star region show it to be surrounded by an HII region but it is unknown whether it is the source of the nebula's ionization.

Our final double star is the famous star **Alpha (α) Centauri**, also known as **Rigel Kent** or **Rigel Kentaurus**. The third-brightest star in the sky, this is in fact part of a triple system, with the two brightest components contributing most of the light, the secondary having a magnitude of 1.24. Shining at a magnitude of –0.4, it is bright enough to be observed during twilight and is a nice object to view as the sky darkens. Some observers have claimed that the star is visible in the daylight with any aperture. The system contains the closest star to the Sun, **Proxima Centauri (V 645)**, which is a very faint red dwarf star and also a flare star with frequent bursts, having a maximum amplitude of around one magnitude. Its current magnitude is 11.01. The group also has a very large proper motion (its apparent motion in relation to the background). Unfortunately, it is too far south to be seen by any northern observer. It is also a very close object, so is a difficult system to resolve. However, using a large telescope the triple system makes a fine spectacle, with contrasting colors of a dazzling yellow for the primary, a dull orange secondary, and a faint weak red for Proxima.

Although there are several fine variable stars in Centaurus, we will look at just a few here. The first is **R Centauri**. This is a Mira-type long-period variable star that during its maximum is a bright red star set in a glittering array of background stars. It has a long period of about 546 days and has the odd property of going through two maxima and minima. These minima can vary from between 9th and 11th magnitude but occasionally it fades to a faint 13th magnitude. Its maxima vary from between 5.3 and 5.6. The second, **T Centuari**, is a semiregular variable star that reaches magnitude 6.58 every 90 days, although it has on occasion reached magnitude 5.5. During the minima it falls to a faint 10th magnitude. This means that it can be observed using nothing more than medium-power binoculars. It can easily be found as it is a nice red color and just west of the triangle of stars formed by **i, g** and **k Centauri**. Finally there is **TW Centauri**, which is a long-period variable star varying from 7.5 to 13th magnitude in about 269 days.

Open star clusters are found in abundance in Centaurus. One of the finest is **NGC 3766** (**Caldwell 97**). It is located in a truly beautiful part of the sky, and scanning the region between **Lambda (λ) Centauri** and **Omicron (o) Centauri** is a delight. The open cluster is a rich system with a broad central condensation. It is about 15 arcminutes across and contains many orange, red, white and yellow stars. Visible in binoculars as a faint hazy patch with a few brighter stars sprinkled in it, in telescopes many of its hundred 8–13th magnitude stars can be seen (see Figure 3.18). This is a glorious object that should be on every observer's list.

Figure 3.18. NGC 3766 (Space Telescope Science Institute, AAO, UK–PPARC, ROE, National Geographic Society, and California Institute of Technology).

Figure 3.19. IC 2948 (IAS).

A small and often passed-over open cluster is **Collinder 249 (Caldwell 100, IC 2944)**. Located just southeast of Lambda (λ) Centauri, this small cluster, ~ 1° in diameter, contains around 30 stars of magnitude 9 and brighter. Surrounding the cluster is the very faint nebulosity **IC 2948** (see Figure 3.19).

An [OIII] filter will be needed to see this object, but it will show some detail. It is believed that many of the hot O-type stars in the cluster are responsible for providing the necessary exciting radiation for the nebula.

First discovered in 1752 is the open cluster **NGC 5281**. This is a bright cluster of around 6th magnitude containing between 40 and 50 stars. Spread over an area of 5 arcminutes, it can easily be seen in binoculars, and is a lovely sight with larger apertures (see Figure 3.20). In its center can be seen a pattern resembling two crossed curving lines of stars tinted bluish, white, pale yellow and orange. A 6.5 magnitude star commands attention in the field of view A much fainter group of stars, **NGC 5269**, can be seen about 15 arcminutes to the west.

One of the few clusters that can be glimpsed without any optical aid, providing the conditions are excellent, is **NGC 5460**. This has the largest apparent diameter of any open cluster in Centaurus, and with small binoculars will reveal itself as a hazy patch of light with a solitary 8th magnitude star (see Star Chart 3.15). In small telescopes it will be about 25 arcminutes in diameter and a semicircle of bright stars can be observed to the north of its center (see Figure 3.21). Incidentally, about 20 arcminutes to the northeast of this semicircle is a small lenticular galaxy, **ESO 221–G26**. With a dark sky and an aperture of about 20 cm, it can easily be seen as a 50 × 30 arcseconds smudge. Its estimated distance is 15 million parsecs.

A very fine open cluster is **NGC 5617**, some 15 arcminutes across, which shows some central condensations. Set in a lovely rich star field, it is a nice object seen even in a small

Figure 3.20. NGC 5281 (Space Telescope Science Institute, AAO, UK–PPARC, ROE, National Geographic Society, and California Institute of Technology).

Star Chart 3.15. NGC 5460, ESO 221–G26.

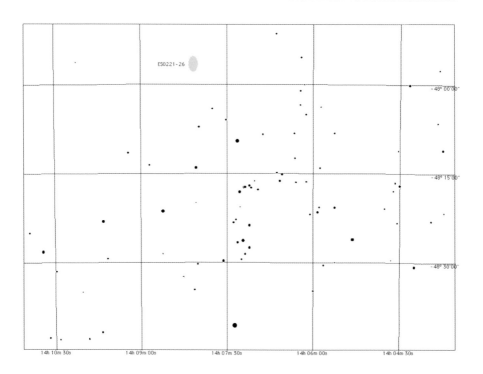

Figure 3.21. NGC 5460 (Space Telescope Science Institute, AAO, UK–PPARC, ROE, National Geographic Society, and California Institute of Technology).

telescope, say 10 cm (see Figure 3.22). Situated about 12 arcminutes southeast of NGC 5617 is the very faint cluster **Pismis 19**. In a telescope of at least 30 cm aperture it will appear as a small 2 arcminute misty glow.

To conclude our tour of open cluster let's look at an object that causes some controversy: **NGC 5299**. This is a large collection of stars ~ 1° in diameter, and can be found

Figure 3.22. NGC 5617 (Space Telescope Science Institute, AAO, UK–PPARC, ROE, National Geographic Society, and California Institute of Technology).

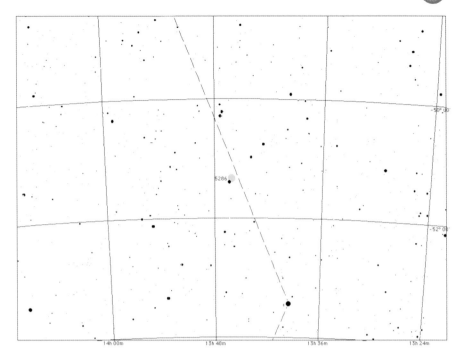

Star Chart 3.16. NGC 5286.

about 2° west of Beta Centauri. It does stand out from the surrounding star field. However, it is now believed that it is just a local concentration of the Milky Way, and not a cluster at all. What do you think? Does it warrant the classification as a cluster?

There are three globular clusters that can be observed in Centaurus, including the most famous and brightest in the entire sky, but more about that later. First let's look at some of its fainter, but equally interesting cousins. Oddly enough, our first globular, **Rup 106**, was actually first cataloged as an open cluster in 1961, by J. Ruprecht. However, it has now been correctly identified and reclassified. It is a faint indeterminate hazy object some 2 arcminutes across. It contains many very faint stars with a few brighter foreground stars superimposed, with others lying to the north and west. You will need at least a 20 cm telescope to observe this elusive cluster.

A much brighter globular cluster of magnitude 7.7 is **NGC 5286** (**Caldwell 84**) (see Star Chart 3.16). This is a nice object to view, but at least a 10 cm telescope will be needed in order to resolve any of the brighter and outlying members of the cluster into stars (see Figure 3.23). The brilliant core will remain, however, an unresolved disk. One problem you may experience, especially in a low-power field, will be the presence of the star **M Centuari**, magnitude 4.64. This deep yellow (some observers prefer to call it orange) star is only 4 arcminutes southeast of the cluster and so its glare could be a problem. It is a spectroscopic binary star of period 437 days. The cluster is believed to be some 9000 parsecs away.

I have purposely left the best till last. First recorded by Ptolemy as a star, and then subsequently by Bayer as **NGC 5139** (**Caldwell 80**), also known as the **Omega (ω) Centauri,** it had to wait until 1677 to be correctly identified by Edmund Halley as a globular cluster. It is without a doubt an amazing object. I recall the first time I saw it, and I was astounded. It

Figure 3.23. NGC 5286 (Space Telescope Science Institute, AAO, UK–PPARC, ROE, National Geographic Society, and California Institute of Technology).

still takes my breath away several years after that first viewing. It is visible with the naked eye as a large fuzzy patch and there are some reports of observers being able to see it as a definite oblate (see Star Chart 3.17). With the naked eye it subtends an angle of around 36 arcminutes (a full moon's width!) and photographs show it to extend to well over 60 arcminutes. In fact individual stars that belong to the cluster have been observed as far out as some 90 arcminutes, which translates into 150 parsecs. With small binoculars it is not resolved yet you feel it nearly is; with large binoculars some of its estimated 1 million stars become resolved. In small telescopes of, say, 8 cm aperture, it starts to get perceptibly grainy, and then in larger telescopes it literally becomes almost unbelievable with structure and definition that defy words. It is one of the most beautiful sights in the entire sky (see Figure 3.24). It will come as a surprise, I am sure, to know that observers as far north as New York, or say Athens, can actually glimpse this gem of an object. However, it will be so close to the horizon that atmospheric interference will make it all but unrecognizable. Nevertheless, try to find it if you have the choice, as once seen it will never be forgotten. Current estimates place it at a distance of about 5200 parsecs from us. Observe this globular whenever you have the chance!

Several planetary nebulae are also visible in Centaurus, and fortunately one that can be seen even in binoculars. Our first is a difficult object and will need at least a 30 cm telescope. This is **Lo 5**, discovered by A. Longmore in 1976. It is a large annular planetary nebula located in a moderately rich star field. It goes without saying that a dark and very transparent sky will be needed, and of course our old friend the [OIII] filter will help considerably and show it to be about 22.5 arcminutes in diameter. A somewhat brighter planetary is **NGC 3699**. It is small at about 1 arcminute across, but it can be seen in a 20 cm telescope and displays definite mottling and structure. If an [OIII] filter is used it can be seen in a smaller telescope of, say, 10 cm aperture.

Discovered in 1834 by John Herschel, **NGC 3918** is a rare type of nebula in that it can be seen with binoculars. However, do not expect much, as all you will see is a point of light (see Star Chart 3.18). But there will be something about it that just doesn't look like a star. It may be that it is not as sharp as the images of the stars, or perhaps its blue color will give it away, and this gives it its name, the **Blue Planetary**. A small telescope of, say, 5 cm will show the disk of about 2 arcseconds diameter. A challenge to large binocular observers is to see if you can locate its 11th magnitude central star. Good luck!

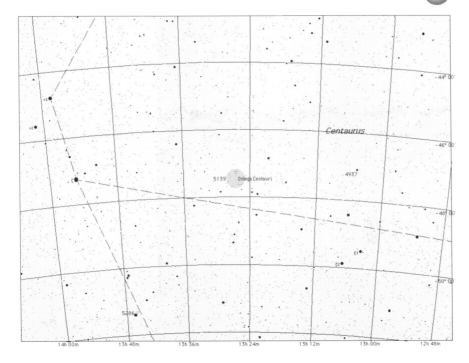

Star Chart 3.17 (*above*). Omega Centauri.
Figure 3.24 (*below*). Omega Centauri (Matt BenDaniel, http://starmatt.com).

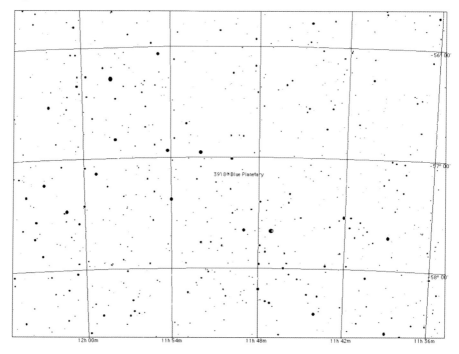

Star Chart 3.18. NGC 3918.

Another faint and recently discovered planetary is **K 1–23**, discovered by Lubos Kohoutek in 1971. It will appear slightly oval in shape, and is just visible in a 20 cm telescope. Set in a rich field of stars is the faint and somewhat hazy planetary **NGC 5307**. It is about 10 arcseconds across and shows a definite elliptical shape. Under good seeing conditions, a small, say 8 cm, telescope will show it as an out-of-focus star.

Our final planetary is quite surprising. It is **PK 315–0.1 (He 2–111)**, and is a small and elongated object located close to the open cluster NGC 5617 (see earlier) and located in the same field of view as Alpha Centauri (see Star Chart 3.19). With an [OIII] filter it shows a size 25 × 15 arcseconds and in a 20 cm telescope will be an easy object. However, it was discovered that deep images of this planetary revealed an enormous filamentary halo surrounding it, making the actual size 10 × 5 arcminutes, thus making it one of the largest planetary nebulae known.

There is one reflection nebula we can observe, and that is **NGC 5367**. It will appear as a faint circularish haze about 2 arcminutes across surrounding the double star **h 4636**, magnitudes 10.3 and 10.7, that are apparently responsible for providing the light which is subsequently reflected (see Star Chart 3.20). Deep imaging reveals that the nebula is actually the head of a cometary globule **CG 12**, and is an active region of star formation (see Figure 3.25).

Our last major topic concerns the many galaxies in Centaurus. We will of course just concern ourselves here with those that reside in the region of the Milky Way, although there are quite a few that do not.

Our first galaxy is the most conspicuous elliptical galaxy in the rich **Centaurus Cluster** of galaxies, **NGC 4696**. It is easily visible in a 15 cm telescope, nearly 2 arcminutes across (see

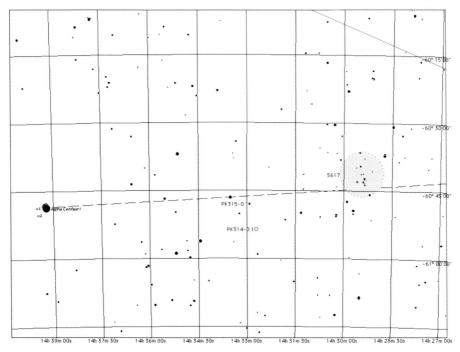

Star Chart 3.19 (*above*). PK 315–0.1.
Star Chart 3.20 (*below*). NGC 5367.

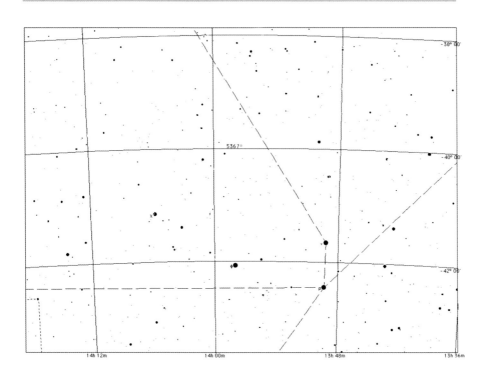

Figure 3.25. NGC 5367 (Space Telescope Science Institute, AAO, UK–PPARC, ROE, National Geographic Society, and California Institute of Technology).

Figure 3.26). Nearly 15 arcminutes west and a little south is the much fainter galaxy **NGC 4709**. This is an elliptical galaxy just spotted in a 15 cm telescope. Those observers with larger telescopes will benefit from scanning this area as there are about 20 other galaxies in this region that can be seen. The distance for this cluster is about 30 million parsecs, or twice that of the northern sky's Virgo cluster (see Star Chart 3.21 for both galaxies).

Figure 3.26. NGC 4696 (Space Telescope Science Institute, AAO, UK–PPARC, ROE, National Geographic Society, and California Institute of Technology).

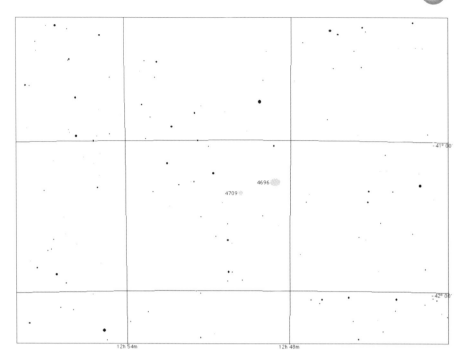

Star Chart 3.21. NGC 4696; NGC 4709.

Set amongst a lovely star field is the spiral galaxy **NGC 4945 (Caldwell 83)**. It will appear as a long hazy object some 20 arcminutes by 4.5 arcminutes (see Star Chart 3.22). Fairly bright in its middle parts, it tends to fade out towards its ends. Its visual magnitude is around 8.5. Binoculars will show it as a tiny sliver of gray light roughly a third of the Moon's diameter (see Figure 3.27). An 8 cm telescope will show it as a 10 arcseconds streak of light. However, its appearance will alter dramatically as one uses larger telescopes. A 25 cm telescope will reveal a tiny bright nucleus surrounded by the faint mottled spiral arms. Due to interstellar absorption by dust, it is believed that the galaxy is in fact 1 magnitude brighter than it appears. It belongs to a small group of galaxies that also contains **NGC 5128** and **M 83**. Lying in the same field as NGC 4945, but unrelated to it, is **NGC 4976**, a nice elliptical galaxy some 2 arcminutes across, and easily seen in a 10 cm telescope. It shows a small but bright nucleus with hazy edges (see Star Chart 3.22).

Our final galaxy is a very unusual object, an active galaxy, **NGC 5128 (Caldwell 77)**. Under excellent seeing conditions it can be seen in finder scopes and small binoculars as a faint 7th magnitude star some 4° north of Omega Centauri, and with large binoculars a tiny matched pair of semicircles separated by a dark band, measuring some 18 arcminutes by 14 arcminutes (see Star Chart 3.23). A 35 cm telescope at medium magnification will show NGC 5128 as a bright circular patch of hazy light, very bright towards its middle, bisected by a dust lane that shows some structure (see Figure 3.28). Also known as **Centaurus A**, research has indicated that in addition to being a very strong radio source with two enormous radio lobes on either side, the curious morphology of NGC 5128 is in fact the result of a merger between two galaxies, a large elliptical and a smaller spiral. Alas, it is too far south for northern observers, but can be seen from the southern USA.

Star Chart 3.22
(*above*). NGC 4945;
NGC 4976.

Figure 3.27 (*left*).
NGC 4945 (Space
Telescope Science
Institute, AAO,
UK–PPARC, ROE,
National Geographic
Society, and California
Institute of Technology).

Our final object that should be included, as it stands out plainly for all to see without any optical aid, is the vast dark cloud silhouetted against the very rich Milky Way. It is about 10 × 4 arcminutes in size, roughly elliptical in shape and with a somewhat irregular edge. It lies roughly north–south. A small telescope will show it easily, as well as the many other smaller and less obvious dark clouds in the same region.

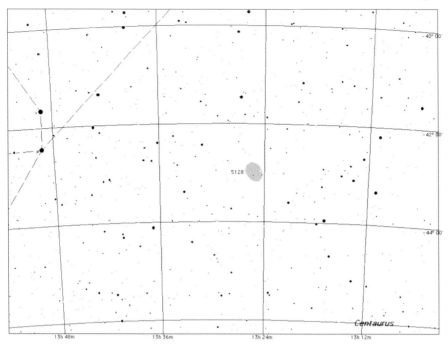

Star Chart 3.23. NGC 5128.

Figure 3.28. NGC 5128 (IAS).

3.5 Circinus

Lying in this section of the Milky Way is the small and fairly obscure collection of stars known as **Circinus**. It culminates at the end of April, and is high in the sky for southern observers (see Star Chart 3.24). It holds little to grab the attention of amateurs, yet it does hold a few delights and one surprise.

Both **Alpha (α) Circini** and **Gamma (γ) Circini** are nice double stars. The former, Alpha Circini, has a bright yellow primary and faint red secondary separated by 15.7 arcseconds. Recent research indicates that the primary is a ACV-type variable star.[7] It can be easily spit by a small telescope of, say, 8 cm. The latter, Gamma Circini, is a nice colorful system, with a pale yellowish 4.94 magnitude star and an 8th magnitude red star some 3 arcseconds to the northeast. These two stars are not really part of a physical system, but **Gamma^A Circini** is itself a binary with an exceedingly close 5.82 magnitude secondary, **Gamma^B Circini**. It is at a separation of about 1 arcsecond, and in a 15 cm telescope, will appear as a slightly elongated star that only becomes fully resolved into two components with apertures of 20 cm and greater.

A third double star is **h (Herschel) 4632**. Set in a lovely star field, this pair of unequal magnitude stars, colored pale yellow and white, can easily be seen in a telescope of 8 cm aperture. About 9 arcseconds southwest of the pair is a nice spiral of stars that is in the same field of view.

Star Chart 3.24. Circinus.

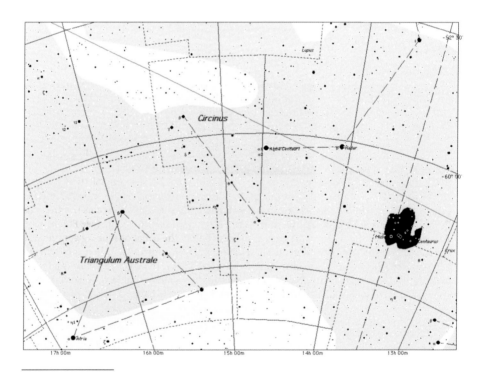

---

[7] Named after the prototype of its class, Alpha (α) Canum Veniticorum, a rotating star.

Figure 3.29. NGC 5823 (Space Telescope Science Institute, AAO, UK–PPARC, ROE, National Geographic Society, and California Institute of Technology).

There are also a couple of variable stars we can look at. **AX Circini** is a Cepheid variable whose magnitude ranges from 5.6 to 6.19 with a period of 5.3 days. **Theta (θ) Circini** is an irregular variable star class GCAS,[8] whose magnitude changes from 5.0 to 5.4.

There is a nice cluster in Circinus: **NGC 5823**. Lying just inside the northern border of the constellation, it is a rich open cluster with about 80 to 100 stars with magnitudes of 10th or fainter, and around 10 arcseconds across (see Figure 3.29). It can be glimpsed with a small telescope but can only be appreciated with an aperture of 30 cm and more.

One lone planetary nebula is available for keen observers: **NGC 5315**. In small telescopes it will only appear as a star, albeit a bluish-green one. With a larger telescope and high magnification, the disk will be seen as an object about 5 arcseconds across. The central star has a magnitude of 11.4. To aid in locating the nebula, there is a pale yellow star some 4 arcseconds to the west.

Our final object should come as a surprise. Considering that the constellation is buried deep in the Milky Way, it is something of a pleasant surprise that we can see another galaxy. **ESO 97–G13** was discovered as recently as 1977. It is also known as the **Circinus Galaxy** and is a spiral galaxy located only some 4° from the galactic plane (see Figure 3.30). This would imply that there must be some sort of gap in the usual obscuring gas and dust in order for the galaxy to be seen (see Star Chart 3.25). It is buried within a very rich star field, and will look like a pale blur some 2 × 1 arcseconds. In larger telescopes a fainter extended halo can be glimpsed. Deep photographs show a dark lane on its central bulge, giving an appearance similar to that of the Black-Eye Galaxy, M64.

---

[8] Named after the prototype of its class, Gamma (γ) Cassiopeiae, which is a "nova-like" eruptive variable star.

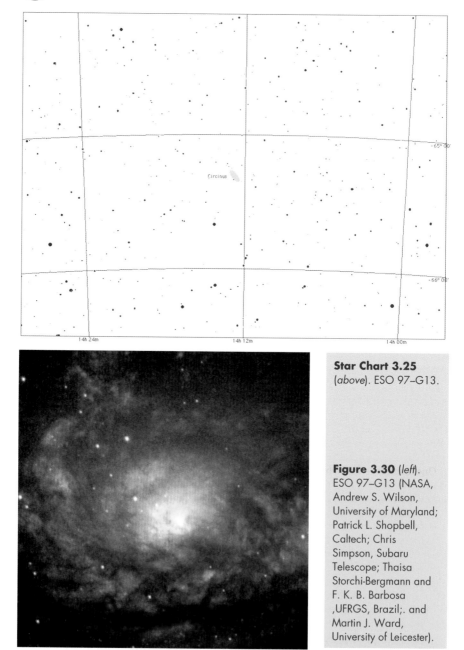

Circinus

-65° 00'

-66° 00'

14h 24m　　　　　　　14h 12m　　　　　　　14h 00m

Star Chart 3.25
(*above*). ESO 97–G13.

Figure 3.30 (*left*).
ESO 97–G13 (NASA,
Andrew S. Wilson,
University of Maryland;
Patrick L. Shopbell,
Caltech; Chris
Simpson, Subaru
Telescope; Thaisa
Storchi-Bergmann and
F. K. B. Barbosa
,UFRGS, Brazil;. and
Martin J. Ward,
University of Leicester).

3.6 Volans

Although this little constellation culminates in January, it is located in, and indeed nearly surrounded by, constellations of March and April; therefore I have decided to include **Volans** here (see Star Chart 3.26). Not that it really matters though, as only its northeast-

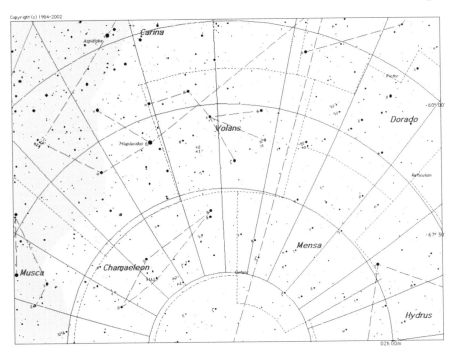

Star Chart 3.26. Volans; Chamaeleon.

ern part is located in the Milky Way, and even then, only one object concerns us. That object is **Epsilon (ε) Volantis**. This is a nice double star, white in color, magnitude 4.37, with a pale yellow companion, magnitude 7.48, well separated and located to the northeast. It can easily be seen in an 8 cm telescope. The primary itself is a spectroscopic binary.

3.7 Octans

Following in the same vein as the previous constellation, **Octans** has little to offer us (see Star Chart 3.27). Only a small part of it resides within the Milky Way. To be fair, it is host to the south celestial pole, but as that is a barren region as well, we have only one object to discuss, the double star **Pi (π) Octantis**. To be truthful this is actually just a couple of stars, π^1 and π^2, which combine into a very wide optical double. Both stars are at magnitude 5.65.

3.8 Chamaeleon

The southern sky, especially around its celestial pole, seems to be in no short supply of fairly obscure constellations,[9] and such a one is **Chamaeleon** (see Star Chart 3.26). Only a

---

[9] In case I get accused of being biased towards the northern sky, let me set the record straight by an example. Who has ever really fully identified and observed Camelopardalis in detail? Not me, that's for sure!

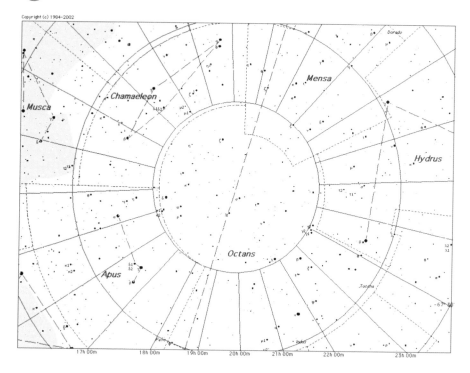

Star Chart 3.27. Octans.

few objects warrant interest here and even though a fair part of the constellation lies within the Milky Way, there is not much to keep us interested. There are two nice double stars. The first is **Delta¹ (δ) Chamaeleontis**. This nice system is in fact the fainter of two stars located about 6 arcminutes apart; the other star is **Delta² (δ) Chamaeleontis**. The magnitudes are 5.5 and 4.45 respectively, colored white and pale yellow. However, Delta¹ Chamaeleontis is in fact a very close double of two stars, magnitudes 6.27 and 6.5 separated by about 1 arcseconds. A 20 cm telescope may be needed in order to resolve it. The second double star is **Epsilon (ε) Chamaeleontis**. This will be a challenge to those amateurs who possess large telescopes, as at least a 25–30 cm aperture will be needed in order to resolve the system. It is a true binary star and it may well be that the star is unresolvable at the present time.

The constellation does have one planetary nebula, **NGC 3195** (see Star Chart 3.28). It is fairly large, around 30 arcseconds and may have a bluish tinge. It can be glimpsed with a 10 cm telescope, but of course a larger aperture will always help.

The final objects we will look at include the reflection nebula **IC 2631**. This is a bright nebula that is part of the enormous **Chamaeleon I Association**. When observed with, say, a 20 cm aperture it will appear as an easily recognizable circular haze about 2 arcminutes across enveloping a 9th magnitude star (see Figure 3.31). It should be visible in a 10 cm telescope, but of course may not be so easily observed. Located about 45 arcminutes to the south is another reflection nebula **Ced 110**, which will require a 30 cm telescope in order to be seen. Within the general area is the very faint spiral galaxy **NGC 3620**, located about 20 arcminutes northwest of IC 2631. Many observers report that this can just be glimpsed in a 20 cm telescope.

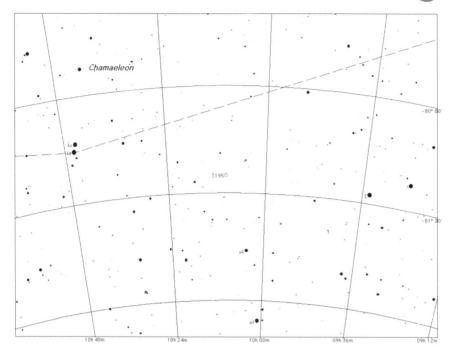

Star Chart 3.28. NGC 3195.

Figure 3.31. IC 2631 (Space Telescope Science Institute, AAO, UK–PPARC, ROE, National Geographic Society, and California Institute of Technology).

3.9 Telescopium

This is the last constellation in this section, and has a few objects that we can observe. **Telescopium** is also one of those constellations that bears no resemblance at all to its namesake (see Star Chart 3.29). How Lacaille in 1752 ever thought he should name this after a telescope is beyond understanding as the brightest stars, forming a small quadrilateral, are in its most northwestern corner. Nevertheless, a telescope it has been named, and indeed we shall need one to see some the objects below. The Milky Way only grazes the constellation, again in the northwestern part.

To start with a nice double star we can look at **Delta (δ) Telescopii**. This is an easy pair to resolve even in binoculars. Shining with magnitudes 4.92 and 5.07, and separated by 9 arcminutes, they shine with a dazzling steely blue glint typical of B-type stars. They are not, however, associated, but just lined up along our line of sight. Another double is **h (Herschel) 5114**, which can also be seen through binoculars. The primary, a K-type star, has a definite orange hue, whilst the secondary is a pale yellow G-type star. Their magnitudes are 5.67 and 8.3, respectively.

An object that will prove to be a challenge is the planetary nebula **IC 4699**. It will need a 30 cm telescope at least, and even then will only appear as a tiny pale disk. There is an 11th magnitude star about 1 arcminute to its north which can be used as a guide if needs be, and a pair of stars 1 arcminute to its southwest (see Star Chart 3.30). The nebula is about 3 arcseconds in diameter.

Star Chart 3.29. Telescopium.

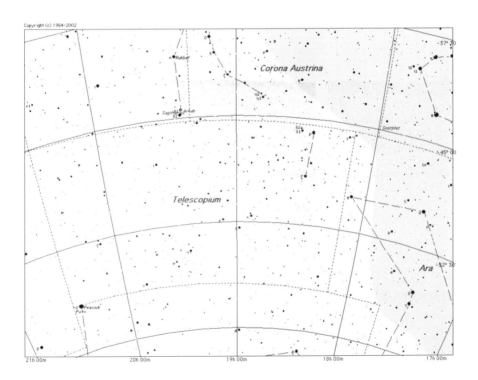

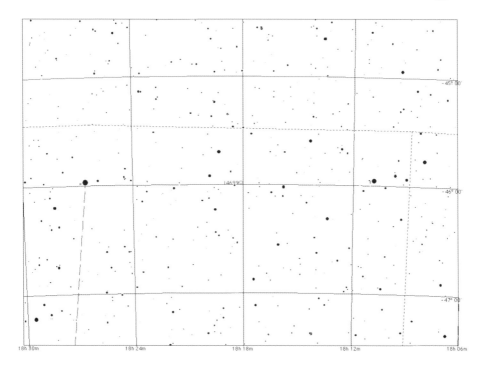

Star Chart 3.30. IC 4699.

A nice little asterism is the group of stars **Harrington 8**. The name was self-penned by the American amateur astronomer Phil Harrington (see Star Chart 3.31). It consists of several 9th magnitude stars located just south of Delta[1] Telescopii. It does look a little like the letter "X", and so has got the nickname the **X Marks The Spot Cluster**. Easily visible in large binoculars and small telescopes.

Our final object is the globular cluster **NGC 6584**. It isn't particularly bright, but can be seen in a 20 cm telescope that will resolve some of its structure (see Star Chart 3.32). It is of the broadly condensed type of globular and has some outlying members scattered irregularly about it. With a diameter of 2.5 arcminutes it can just be glimpsed with a 10 cm telescope, although it will only then appear as a faintly hazy object (see Figure 3.32).

The following constellations are also visible during these months at different times throughout the night. Remember that they may be low down and so diminished by the effects of the atmosphere. Also, you may have to observe them either earlier than midnight, or some considerable time after midnight, in order to view them.

Northern Hemisphere

Antila, Auriga, Camelopardalis, Canis Major, Canis Minor, Cassiopeia, Cepheus, Corona Austrini, Cygnus, Gemini, Lacerta, Lupus, Lyra, Monoceros, Ophiuchus, Orion, Perseus, Puppis, Pyxis, Sagitta, Sagittarius, Scorpius, Scutum, Serpens, Vulpecula.

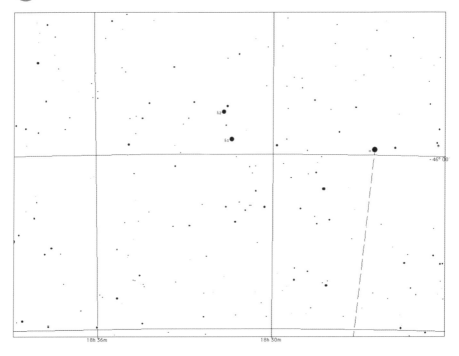

Star Chart 3.31 (*above*). Harrington 8.
Star Chart 3.32 (*below*). NGC 6584.

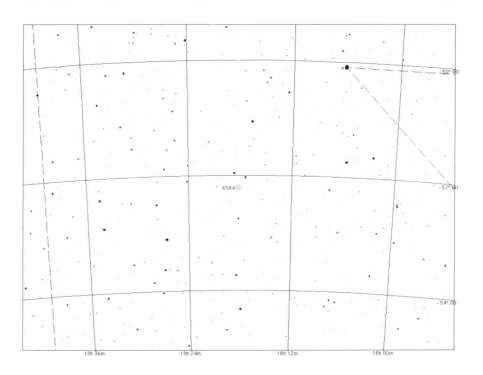

Figure 3.32. NGC 6584 (Space Telescope Science Institute, AAO, UK–PPARC, ROE, National Geographic Society, and California Institute of Technology).

Southern Hemisphere

Antila, Apus, Aquila, Ara, Canis Major, Canis Minor, Columba, Corona Australis, Delphinus, Gemini, Hercules, Hydra, Lepus, Libra, Lupus, Monoceros, Norma, Ophiuchus, Orion, Pavo, Puppis, Pyxis, Sagitta, Sagittarius, Scorpius, Scutum, Serpens Cauda, Taurus, Triangulum Australe, Vela, Vulpecula.

Objects in Carina

Stars

Designation	Alternate name	Vis. mag	RA	Dec.	Description
h 4130	Herschel 4130	6.5, 4.3	$08^h\,40.7^m$	-57° 33'	PA 244°; Sep. 3.6"
Upsilon (υ) Carinæ	Rmk 11	3.0, 6.2	$09^h\,47.2^m$	-65° 04'	PA 127°; Sep. 5.0"
Hrg 47	Hargreave 47	6.4, 7.9	$10^h\,03.6^m$	-61° 53'	PA 351°; Sep. 1.2"
t² Carinæ	Dunlop 94	4.7, 8.1	$10^h\,38.7^m$	-59° 11'	PA 211°; Sep. 14.5"
Gli 152		6.0, 8.8	$10^h\,39.0^m$	-58° 49'	PA 79°; Sep. 25"
R 164		6.2, 9.7	$10^h\,59.2^m$	-61° 19'	PA 76°; Sep. 3.9"
R Carinæ	HD 82901	6.10	$09^h\,32.2^m$	-62° 47'	Variable star
S Carinæ	HD 88366	5.6	$10^h\,09.4^m$	-61° 33'	Variable star

Deep-Sky Objects

Designation	Alternate name	Vis. mag	RA	Dec.	Description
IC 2220	Toby Jug	–	$07^h\,56.8^m$	-59° 07'	Bipolar reflection nebula
NGC 3503			$11^h\,01.3^m$	-59° 51'	Reflection nebula
NGC 3199		–	$10^h\,17.1^m$	-57° 55'	Nebula and Wolf–Rayet star
NGC 3247	Westerlund 2	10.5	$10^h\,23.9^m$	-57° 45'	Cluster and nebulosity
NGC 3324		–	$10^h\,37.3^m$	-58° 38'	Diffuse nebula
NGC 3581–82			$11^h\,12.3^m$	-61° 16'	Emission nebula
NGC 3603		8.5	$11^h\,15.1^m$	-61° 15'	Open star cluster & nebulosity
NGC 3372		2.5	$10^h\,45.1^m$	-59° 52'	Nebula around Eta (η) Carinæ
IC 2448	Caldwell 92	10.4	$09^h\,07.1^m$	-69° 57'	Planetary nebula
NGC 2867	Caldwell 90	9.7	$09^h\,21.4^m$	-58° 19'	Planetary nebula
IC 2501	PK 281–5.1	10.4	$09^h\,38.8^m$	-60° 05'	Planetary nebula
IC 2553	PK 285–05.1	10.3	$10^h\,09.3^m$	-62° 37'	Planetary nebula
NGC 3211	PK 286–04.1	10.7	$10^h\,17.8^m$	-62° 40'	Planetary nebula
IC 2621	PK 291–4.1	11.2	$11^h\,00.3^m$	-65° 15'	Planetary nebula
NGC 2516	Caldwell 96	3.3	$07^h\,58.3^m$	-60° 52'	Open cluster

Designation	Alternate name	Vis. mag	RA	Dec.	Description
NGC 3114		4.2	10^{h}02.7^m	–60° 07'	Open cluster
NGC 3293	Gem Cluster	4.7	10^{h}35.8^m	–58° 14'	Open cluster
IC 2602		1.9	10^{h}43.2^m	–64° 24'	Southern Pleiades open cluster
NGC 3532	Caldwell 102	3.1	11^{h}06.4^m	–58° 40'	Open cluster
NGC 2808	Caldwell 91	4.62	09^{h}12.0^m	–64° 52'	Globular cluster

Objects in Crux

Stars

Designation	Alternate name	Vis. mag	RA	Dec.	Description
Alpha (α) Crucis	Acrux	1.25, 1.64	12^{h}26.6^m	–63° 06'	PA 111°; Sep. 4.1"
Beta (β) Crucis	Mimosa	1.25, 11.4	12^{h}47.7^m	–59° 41'	PA 321°; Sep. 44"
Gamma (γ) Crucis	Gacrux	1.59, 6.42	12^{h}31.2^m	–57° 07'	Wide double
Mu (μ) Crucis	Dunlop 126	4.03, 5.08	12^{h}54.6^m	–57° 11'	PA 17°; Sep. 45"
R Crucis		6.89	12^{h}23.6^m	–61° 38'	Cepheid variable star

Deep-Sky Objects

Designation	Alternate name	Vis. mag	RA	Dec.	Description
NGC 4755	Caldwell 94	4.2	12^{h}53.6^m	–60° 15'	Open cluster
NGC 4103		7.4p	12^{h}06.7^m	–61° 15'	Open cluster
NGC 4349		7.35	12^{h}24.5^m	–61° 54'	Open cluster
NGC 4052	Coalsack Cluster	6.9	12^{h}42.3^m	–62° 58'	Open cluster
NGC 4609	Caldwell 98	6.89	12^{h}23.6^m	–61° 3'	Open cluster
Ruprecht 98		6.9	11^{h}58.0^m	–64° 29'	Open cluster
Caldwell 99	Coalsack	–	12^{h}53.0^m	–63° 00'	Dark nebula

Objects in Musca

Stars

Designation	Alternate name	Vis. mag	RA	Dec.	Description
h 4432	Herschel 4432	5.36, 6.68	$11^h 23.4^m$	$-67° 57'$	PA 306°; Sep. 2.4"
Beta (β) Muscae	R 207	3.51, 4.01	$12^h 46.3^m$	$-68° 06'$	PA 35°; Sep. 11.2"
Theta (θ) Muscae	Rmk 16	5.64, 7.66	$13^h 08.1^m$	$-65° 18'$	PA 187°; Sep. 5.3"

Deep-Sky Objects

Designation	Alternate name	Vis. mag	RA	Dec.	Description
NGC 4463		7.2	$12^h 30.0^m$	$-64° 48'$	Open cluster
NGC 4815		8.6	$12^h 58.0^m$	$-64° 57'$	Open cluster
NGC 4372		7.8	$12^h 25.8^m$	$-72° 40'$	Open cluster
NGC 4833		7.4	$12^h 59.6^m$	$-70° 53'$	Open cluster
NGC 5189	Spiral Planetary	10.3p	$13^h 33.5^m$	$-65° 59'$	Planetary nebula
NGC 4071	PK 298–04.1	12.8	$12^h 04.2^m$	$-67° 18'$	Planetary nebula
IC 4191	PK 304–4.1	10.6	$13^h 08.8^m$	$-67° 139'$	Planetary nebula

Objects in Centaurus

Stars

Designation	Alternate name	Vis. mag	RA	Dec.	Description
h 4423	Herschel 4423	6.9, 7.3	$11^h16.5^m$	$-45°\,53'$	PA 275°; Sep. 2.5"
Brs 5		7.8, 8.6	$11^h24.7^m$	$-61°\,39'$	PA 238°; Sep. 5.9"
D Centauri	Rmk 14	5.75, 6.98	$12^h14.0^m$	$-45°\,43'$	PA 243°; Sep. 2.8"
Gamma (γ) Centauri	Muhlifain	2.85, 2.95	$12^h41.5^m$	$-48°\,58'$	Sep. 0.2"–1.7"
I 424		4.5ᵥ, 8.4	$13^h12.3^m$	$-59°\,55'$	PA 7°; Sep. 1.9"
Δ 141	Q Cen	5.24, 6.63	$13^h41.7^m$	$-54°\,133'$	PA 163°; Sep. 5.4"
R 223	Russell 223	6.5, 10.0	$13^h38.1^m$	$-58°\,25'$	PA 23°; Sep. 2.5"
CorO 157	V766 Cen	6.2–7.5, 9.9	$13^h47.2^m$	$-62°\,35'$	PA 318°; Sep. 9.4"
Howe 24		6.6, 10.2	$13^h48.9^m$	$-35°\,42'$	PA 355°; Sep. 11.6"
Δ 148	3 Centauri	4.51, 6.05	$13^h51.8^m$	$-33°\,00'$	PA 106°; Sep. 18.1"
Beta (β) Centauri	Hadar	0.58, 3.95	$14^h03.8^m$	$-60°\,22'$	PA 251°; Sep. 1.3"
Alpha (α) Centauri	Rigel Kent	0.14, 1.24	$14^h39.6^m$	$-60°\,50'$	Sep. 1.7"–21.7"
R Centauri		5.30	$14^h16.6^m$	$-59°\,55'$	Variable star
T Centauri		6.58	$13^h41.8^m$	$-33°\,35'$	Variable star
TW Centauri		8.8–12.6p	$13^h57.7^m$	$-31°\,04$	Variable star

Deep-Sky Objects

Designation	Alternate name	Vis. mag	RA	Dec.	Description
NGC 3766	Caldwell 97	5.3	$11^h36.1^m$	$-61°\,37'$	Open cluster
Collinder 249	Caldwell 100	4.5	$11^h36.6^m$	$-63°\,02'$	Open cluster
NGC 5281		5.9	$13^h46.6^m$	$-62°\,54'$	Open cluster
NGC 5460		5.6	$14^h07.6^m$	$-48°\,19'$	Open cluster
NGC 5617		6.3	$14^h29.8^m$	$-60°\,43'$	Open cluster
NGC 5299		–	$13^h50.3^m$	$-59°\,52'$	Open cluster
Rup 106		11.0	$12^h38.7^m$	$-51°\,09'$	Globular cluster
NGC 5286	Caldwell 84	7.6	$13^h46.4^m$	$-51°\,22'$	Globular cluster
NGC 5139	Caldwell 80	3.7	$13^h26.8^m$	$-47°\,28'$	Globular cluster

Designation	Alternate name	Vis. mag	RA	Dec.	Description
Lo 5	PK 286+11.1	12.5	11h13.8m	−48° 06'	Planetary nebula
NGC 3699	PK 292+1.1	11.3	11h28.0m	−59° 57'	Planetary nebula
NGC 3918	Blue Planetary	8.4p	11h50.3m	−57° 11'	Planetary nebula
K 1−23	PK 299+18.1	13.0	12h30.9m	−44° 14'	Planetary nebula
NGC 5307	PK 312+10.1	11.2	13h51.1m	−51° 12'	Planetary nebula
He 2−111	PK 315−0.1	11.3	14h33.3m	−60° 50'	Planetary nebula
IC 2948			11h38.3m	−63° 22'	Emission nebula
NGC 5367		−	13h57.7m	−39° 59'	Reflection nebula
NGC 4696	Caldwell 83	10.2	12h48.8m	−41° 19'	Galaxy
NGC 4945		8.6	13h05.4m	−49° 28'	Galaxy
NGC 4976		10.2	13h08.6m	−49° 30'	Galaxy
NGC 5128	Caldwell 77	6.4	13h25.5m	−43° 01'	Galaxy

Objects in Circinus

Stars

Designation	Alternate name	Vis. mag	RA	Dec.	Description
Alpha (α) Circini	Dunlop 166	3.18, 8.57	14h42.5m	−64° 59'	PA 228°: Sep. 15.6"
Gamma (γ) Circini	Herschel 4757	4.94, 5.82	15h23.4m	−59° 19'	PA 18°: Sep. 0.83"
h 4632	Herschel 4632	6.2, 10.0	13h58.5m	−65° 48'	PA 14°: Sep. 6.4"
AX Circini		5.6–6.19	14h52.6m	−63° 48'	Variable star
Theta (θ) Circini		5.36	14h56.8m	−62° 46'	Variable star

Deep-Sky Objects

Designation	Alternate name	Vis. mag	RA	Dec.	Description
NGC 5823		7.9	15h05.7m	−55° 36'	Open cluster
NGC 5315		9.8	13h54.0m	−66° 31'	Planetary nebula
ESO 97−G13	Circinus Galaxy	10.0	14h13.2m	−65° 20'	Galaxy

Objects in Volans

Stars

Designation	Alternate name	Vis. mag	RA	Dec.	Description
Epsilon (ε) Volantis	Rmk 7	4.37, 7.48	08^h 07.9^m	–68° 37'	PA 23°; Sep. 5.4"

Objects in Octans

Stars

Designation	Alternate name	Vis. mag	RA	Dec.	Description
Pi (π) Octantis		5.65, 5.65	15^h 01.8^m	–83° 14'	Wide double star

Objects in Chamaeleon

Stars

Designation	Alternate name	Vis. mag	RA	Dec.	Description
Delta^1 (δ) Chamaeleontis	I 294	6.27, 6.50	10^h 45.2^m	–80° 28'	PA 84°; Sep. 0.78"
Epsilon (ε) Chamaeleontis	Herschel 4486	5.32, 6.06	11^h 59.6^m	–78° 13'	PA 203°; Sep. 0.43"

Deep-Sky Objects

Designation	Alternate name	Vis. mag	RA	Dec.	Description
NGC 3195	PK 296–20.1	11.6	10^h 09.5^m	–80° 52'	Planetary nebula
IC 2631	Ced 112	–	11^h 09.8^m	–76° 37'	Reflection nebula

Objects in Telescopium

Stars

Designation	Alternate name	Vis. mag	RA	Dec.	Description
Delta (δ) Telescopii		4.92, 5.07	$18^h31.8^m$	−45° 55'	Wide double star
h 5114	Herschel 5114	4.92, 5.07	$19^h27.8^m$	−54° 20'	Wide double star

Deep-Sky Objects

Designation	Alternate name	Vis. mag	RA	Dec.	Description
IC 4699	PK 348−13.1	13.3	$18^h18.5^m$	−45° 59'	Planetary nebula
Harrington 8	X-Marks the Spot	−	$18^h30.4^m$	−46° 08'	Asterism
NGC 6584		9.2	$18^h18.6^m$	−52° 13'	Globular cluster

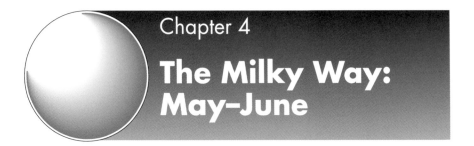

Chapter 4

The Milky Way: May–June

Triangulum Australe, Lupus, Norma, Ara, Libra, Pavo, Apus, Scorpius, Ophiuchus, Corona Australis.

R.A 14^h to 18^h; Dec. $-70°$ to $-30°$; Galactic longitude[1] $325°$ to $10°$; Star Chart 4

4.1 Triangulum Australe

We are now starting to leave those constellations which are the sole domain of the southern hemisphere, and are slowly beginning to visit areas of the sky that can be enjoyed, if not by all amateurs, then most. In addition, we will be looking at not only what some would say are the most spectacular regions of the Milky Way, but also towards the central regions of our own Galaxy (see Star Chart 4).

We start with a very modest constellation, containing very few objects, and end up with a constellation that many think is one of the most richly populated in the Milky Way from an amateur astronomy point of view.

Our first constellation is the small **Triangulum Australe** (see Star Chart 4.1). It is easy to find due to its easily recognizable triangle of three brightish stars. It transits in the middle of May and is circumpolar for most southern observers. It is totally within the Milky Way, situated along its southernmost borders. It is the perfect type of constellation for sweeping with binoculars, as the star fields are, as would be expected, full of fine vistas of star-strewn fields of view. However, it has few deep-sky objects: one cluster, a few planetary nebulae, several very faint galaxies (which we will not discuss), and some double and variable stars.

Let's star by looking at the variable stars. There are three Cepheid-type variable stars that are visible in binoculars and small telescopes, and so would make ideal observing projects. The first, **R Trianguli Australis**, varies in magnitude from 6.4 to 6.9 in a well-defined period of 3.389 days. **S Trianguli Australis** is a nice deep red star and varies over 6.323 days, changing its magnitude from 6.1 to 6.8. Our final variable is **U Trianguli Australis**. This varies over 2.568 days, and is somewhat fainter than the preceding two at magnitude 7.5 to 8.3.

Several double stars are also available to us. Set amidst a profuse star field is a bright star with a very close companion, **I 332**, that will need a telescope of at least 20 cm in order

---

[1] See Appendix 1 for details on astronomical coordinate systems.

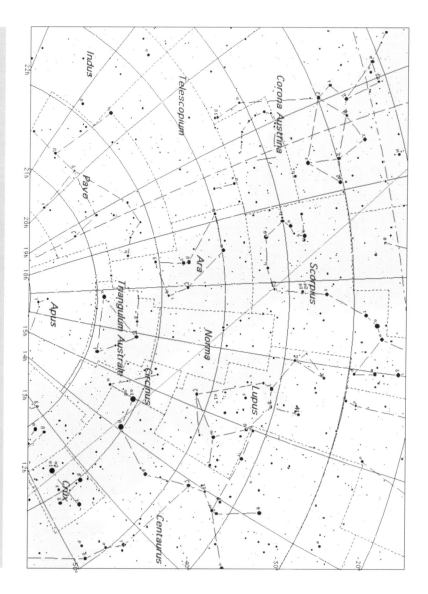

Star Chart 4. May–June.

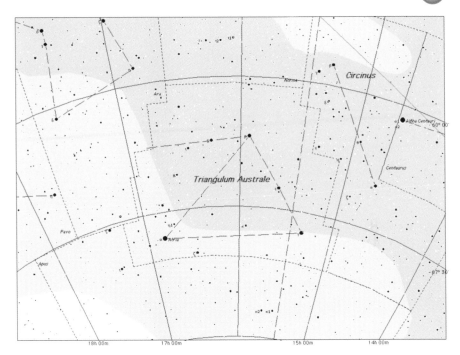

Star Chart 4.1. Triangulum Australe.

to split the system. Then there is **Slr 11**, another white star with a close companion, but this time a smaller telescope of, say, 10 cm aperture will be sufficient to resolve this particular double star, as they are of magnitudes 6.5 and 8.5 separated by a close 1.2 arcseconds. There are also two nice wide companion stars to the northeast and southwest, of 9th and 8th magnitude, separated by 45 and 48 arcseconds respectively. It is not known if these are all connected or just stars lying along the line of view. An interesting double star to observe is **Rmk 20**. It is, to be perfectly honest about it, nothing special, just two stars of nearly equal magnitude located in a very rich star field that can be seen with, say, an 8 cm telescope. However, what makes it so interesting is that to the north and south of the pair the field seems particularly empty of stars. This would indicate the presence of dark nebulae. It is a subtle, but real effect. Observe this and decide for yourself.

Our final double star is perhaps the finest in the constellation. It is **Iota (ι) Trianguli Australis (Δ 201)**. It has a nice color contrast of yellow and white with magnitudes 5.28 and 10.3 and lies in a fine field full of stars. The separation, which was 17 arcseconds in the last century, is now decreasing and so may make resolution more and more of a problem. It should, however, for the time being at least, be split in an 8 cm telescope. The primary is also a spectroscopic binary star.

There are two planetary nebulae visible in Triangulum Australe. One is relatively easy to spot, the other less so. **NGC 5979** can be seen, albeit as a slightly out-of-focus star, in an 8 cm telescope (see Star Chart 4.2). With an aperture of, say, 10 cm, a slightly fuzzy circular object will be resolved. Of course, in larger apertures still, the nebula will show as a pale bluish-gray disk some 15 arcseconds wide (see Figure 4.1). Some observers

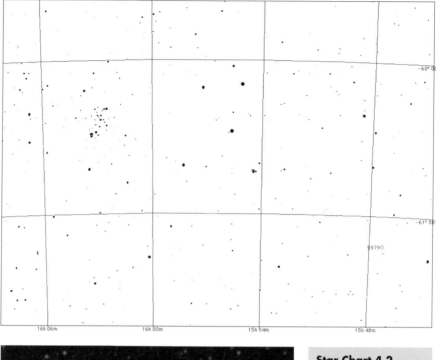

Star Chart 4.2
(*above*). NGC 5979;
NGC 6025.

Figure 4.1 (*left*).
NGC 5979 (Space
Telescope Science
Institute, AAO,
UK–PPARC, ROE,
National Geographic
Society, and California
Institute of Technology).

report that the shape is more elliptical than circular, but all agree that no central star is visible.[2]

One nebula that is almost unknown to amateurs is **NGC 5844**. It will usually appear as an irregularly shaped oval object about 72×50 arcseconds in a 20 cm telescope (see Figure 4.2). Again, as in NGC 5979, no central star can be seen. The use of an [OIII] filter will improve the chances of spotting this one.

---

[2] No doubt this will soon change with more and more astronomers availing themselves of very large telescopes.

Figure 4.2. NGC 5844 (Space Telescope Science Institute, AAO, UK–PPARC, ROE, National Geographic Society, and California Institute of Technology).

Our final object is the open cluster **NGC 6025 (Caldwell 95)**. It is a fine object, visible under excellent conditions to the naked eye, some 15 arcminutes in size (see Star Chart 4.2). In binoculars it will appear as a swarm of thirty 9th magnitude white and pale yellow stars with a bright 7th magnitude star set amongst them. Also, due to the other much fainter stars that are unresolved, it will appear that the cluster is immersed in a faint nebulous haze (see Figure 4.3). A pretty cluster well worth seeking out.

Figure 4.3. NGC 6025 (Space Telescope Science Institute, AAO, UK–PPARC, ROE, National Geographic Society, and California Institute of Technology).

4.2 Apus

This constellation, **Apus**, lies due south of Triangulum Australe only 20° from the southern celestial pole, and so is only visible from southern latitudes (see Star Chart 4.3). Most, although not all, of it lies within the Milky Way. It transits around the latter part of May and has a few objects that demand our attention.

One interesting aspect of Apus that is beyond the reach of amateurs,[3] is the very large refection nebula which covers several degrees of sky in this area, east of **Beta (β)** and **Gamma (γ) Apodis**. This "galactic cirrus" as it has been called covers a large section of the sky towards the southern celestial pole and does in fact reflect the light that is emitted from the Milky Way's disk. But let's now devote our energies to those objects we can see.

There are only a few double stars we can observe, but the first of these, **Delta (δ) Apodis**, is a very easy object. It consists of a very wide pair of ruddy colored stars, separated by over 100 arcseconds with magnitudes of 4.68 and 5.27. They can be split in even the smallest optical equipment, which will clearly show their strong colors. It seems to me that this is a nice object to show to nonamateur astronomers. A much closer pair is **I 236**, separated by only about 2 arcseconds. They should be easily resolved in an 8 cm telescope, which will show a nice bright golden-yellow star with a faint white companion. Let's finish our trawl of double stars with quite a difficult one to resolve, **Cap O 15**. It is relatively easy to locate this star as the field stars are fainter by comparison. It will appear as a bright, lemon-

Star Chart 4.3. Apus.

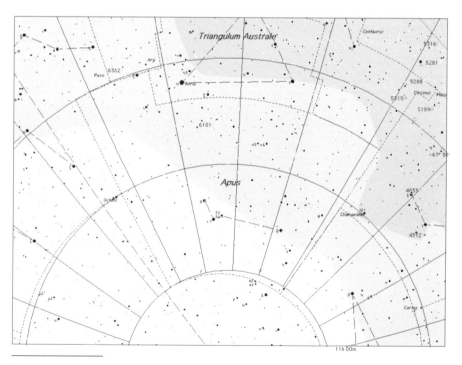

---

[3] It may be that some amateurs equipped with large telescopes and CCDs will be able to image this, and may have already done so!

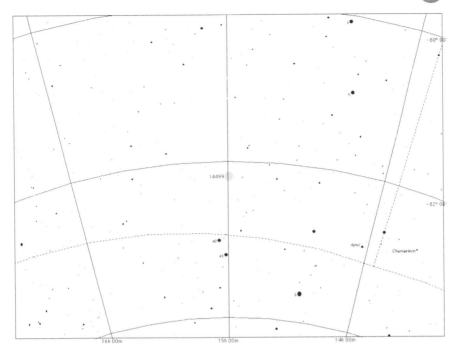

Star Chart 4.4. IC 4499.

yellow star with a very close white companion to its northeast. It should be seen in an 8 cm telescope, and of course larger apertures will make it easier still. There is a nice orange star, **HD 132675**, about 5.5 arcminutes west. If the stars do not hold you enthralled, there is a very faint galaxy, **NGC 5799**, some 25 arcminutes to the south, but be warned, this will only be visible to those who have telescopes with apertures of 30 cm or more.

There is one variable star that may be of interest to you: **Theta (θ) Apodis.** This is a semiregular variable star, and varies in magnitude from 5 to nearly 8th, over a period of about 118 days. This means that it is well within the limits necessary for observation by binoculars. It has a nice red tint, so typical of the M-type star of which it is a member.

Two globular clusters are located within Apus, one easy to observe, the other more of challenge. The former is **IC 4499** (see Star Chart 4.4). It is a faint irregularly shaped cluster about 3 arcminutes across with a magnitude of about 11. Under good seeing conditions, along with a dark sky, you should be able to just glimpse it with a 15 cm telescope, but it won't be easy. It is easier with a telescope of 30 cm but will appear as a hazy object with not much stellar resolution (see Figure 4.4). Nevertheless it is worth seeking out.

The latter cluster is **NGC 6101**, which stands out in contrast with a nice star field (see Figure 4.5). It is admittedly faint at about 9th magnitude, but rich. Again, like IC 4499, it is has an irregular profile, but it does brighten and condense towards its center. In a 10 cm telescope it will appear as a faint hazy unresolved blob, but will be resolved with a 20 cm telescope (see Star Chart 4.5).

Our final object is the planetary nebula **PK 315–13.1 (He 2–131).** This is a small, 10 arc-second diameter object with a bluish color. With a 10 cm telescope it will appear stellar-like but larger apertures will show its nonstellar appearance and under very good conditions a central star may be glimpsed (see Star Chart 4.6). It is unusual in that it has a

Figure 4.4. IC 4499 (Space Telescope Science Institute, AAO, UK–PPARC, ROE, National Geographic Society, and California Institute of Technology).

Star Chart 4.5. NGC 6101.

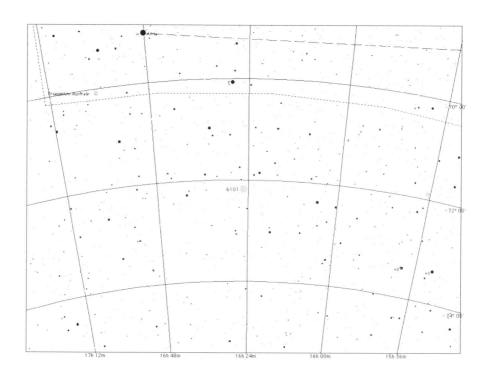

Figure 4.5. NGC 6101 (Space Telescope Science Institute, AAO, UK–PPARC, ROE, National Geographic Society, and California Institute of Technology).

Star Chart 4.6. PK 315–13.1.

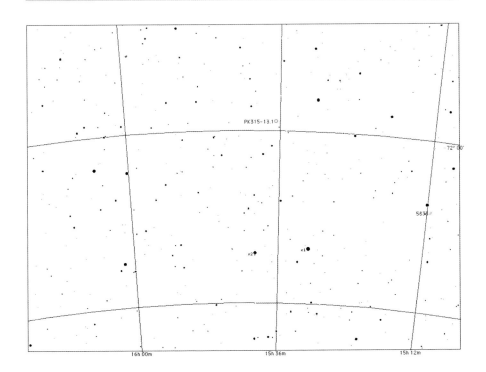

weak [OIII] emission spectrum that is similar to that of **PK 064+05.1**, also known as **Campbell's Star**, in Cygnus.

4.3 Lupus

This is one of those constellations in the Milky Way that has a fine assortment of objects – double stars, open and globular clusters and nebulae of all sorts, yet remains relatively unknown to northern observers. To be fair, **Lupus** is low down for observers in northern Europe, as its northern border is only around 4° below Antares, but it should be adequately visible for amateurs in, say, New York, Miami or Arizona. Nevertheless, it does contain several objects worth searching out, and is well worth scanning with binoculars on dark and clear nights for its almost-resolvable star fields and dark clouds that obscure most of its galaxies (see Star Chart 4.7). Its midnight transit is in early May.

Before we start to look at any single object in detail, it may be interesting to know that several of the brighter stars in this area of the Milky Way are in fact all members of a group of stars that are all moving in the same direction called the **Scorpius–Centaurus Association**. A stellar association is a loosely bound group of very young stars. They may still be swathed in the dust and gas cloud they formed from and star formation will still be occurring within the cloud. Where they differ from open clusters is in the fact that they are enormous, covering both a sizable angular area of the night sky and at the same time encompassing a comparably large volume in space. As an illustration of this huge size, the

Star Chart 4.7. Lupus.

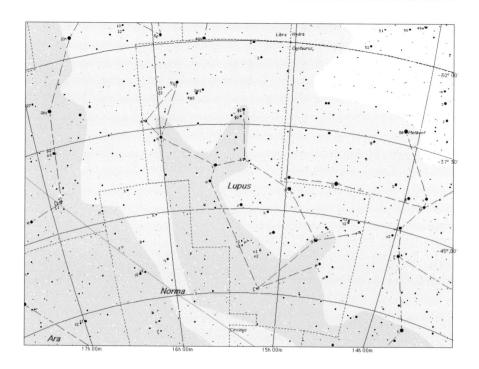

Scorpius–Centaurus Association is around 700×760 light years in extent, and covers about 80°.

It includes most of the stars of 1st, 2nd and 3rd magnitude in Scorpius down through Lupus and Centaurus to Crux. Classed as a B-type association because it lacks O-type stars, its angular size on the sky is around 80°. The center of the association is midway between **Alpha (α) Lupi** and **Zeta (ζ) Centauri**, and its elongated shape is thought to be the result of rotational stresses induced by its rotation about the galactic center. Bright stars in this association include **Theta (θ) Ophiuchi, Beta (β), Nu (ν), Delta (δ)** and **Sigma (σ) Scorpii, Alpha (α)** and **Gamma (γ) Lupi, Delta (δ), Mu (μ)** and **Epsilon (ϵ) Centauri** and **Beta (β) Crucis.** It is a much older, but closer association than the Orion Association.[4]

There are many fine double and multiple stars in Lupus, so let's begin with one that shows a nice color contrast, **h (Herschel) 4690.** This is an easily split system that appear to have colors of orange and blue, although the blue tint may be just a contrast effect with the bright primary. Nevertheless it is a nice sight and should be seen in an 8 cm telescope. A system that will present quite a challenge is **Alpha (α) Lupi.** It has a 2nd magnitude primary and a faint and very close secondary. It may require a 30 cm telescope, or even larger to be fully resolved. Incidentally, the primary is a variable star of the β Cma type, with a small amplitude and period of 0.26 days.

Another fine double is the beautiful system **Pi (π) Lupi.** It is a wide double, and so can be seen in almost any optical equipment. The stars, both a nice pale yellow, have magnitudes of 4.57 and 4.65 and the separation seems to be widening. Both stars are also spectroscopic binaries. Another object that is perfect for all telescopes and set amidst a lovely star field is **Kappa (κ) Lupi (Δ 177).** It consists of a pair of pale yellow stars, magnitudes 3.88 and 5.7. As there has been no apparent change in its separation for a long time now, it can be inferred that this is a physical system in an enormous orbit. A triple system can also make a nice target, and such an example is **Mu (μ) Lupi.** The close pair of the system, separated by about 1.2 arcseconds, can just be split with a 10 cm telescope. The third member is an easy 23 arcseconds away. Their magnitudes are 4.92, 5.50 and 6.87, respectively. Although all the stars are white in color, some observers see a faint red tinge to the third and faintest member of the system. Do you see this color?

An interesting system that is well worth observing is **Gamma (γ) Lupi.** It is a very close binary consisting of two white stars of magnitudes 3.40 and 3.51. What is unusual, however, is that its orbit is almost edgewise on to us, and so the pair will seem to coalesce at regular intervals. The maximum separation will be in 2014 at about 0.8 arcseconds. In a 30 cm telescope this should be resolvable, and in smaller telescopes the image would appear as an elongated star!

Another fine colored pair is **h (Herschel) 4788,** consisting of pale and deep yellow stars. The magnitudes of the stars are 4.66 and 6.62, separated by 3.1 arcseconds. The primary is a possible spectroscopic binary star. Our last double star, but not last multiple, is **Xi (ξ) Lupi.** This is lovely object in any telescope and consists of two almost equal magnitude stars, 5.14 and 5.59, both yellow.

Our final star is **Tau (τ) Lupi.** This is a very wide optical double system of white stars with τ^1 shining at 4.56 and τ^2 at 5.55 magnitude. This is easily seen in small binoculars, but when larger binoculars or a telescope are used, you may just glimpse a much fainter 9th magnitude star between τ^1 and τ^2 that makes it look like a triple star system. However, recent work suggests that although the fainter star is associated with τ^1, neither it nor τ^1 are associated with τ^2.

Let's now start to look at some nebulae, and we will begin with the planetary nebulae. Our first object is a fairly brightish object, **IC 4406.** It is around 28 arcseconds in diameter,

---

[4] This is discussed in Book 1.

Figure 4.6. IC 4406 (Space Telescope Science Institute, AAO, UK–PPARC, ROE, National Geographic Society, and California Institute of Technology).

and appears to some observers with a bluish-gray tint (see Figure 4.6). It should be resolvable with a 10 cm telescope as a tiny hazy cloud of gray light.

However, as you increase both aperture and magnification it will become apparent that the shape of IC 4406 changes from an oval into a definite rectangular shape. As it has a photographic magnitude of 10.6, visually it will have a different magnitude (see Star Chart 4.8). The central star is about 14.7 magnitude and so will be very difficult to see.

A tiny bluish planetary nebula that may need some care in order to be found and resolved is **NGC 5873**. It is only 3 arcseconds across with magnitude 13.5, so will be a challenge. It does, however, lie amongst three bright stars shaped like a triangle which could aid recognition. A much easier and larger planetary to observe is **NGC 5882**. Lying 10° east and about 2.5° south of IC 4406, this is over 7 arcseconds across and can be spotted with ease amongst a nice star field (see Star Chart 4.8). A good way to locate the object is to use a technique used by experienced amateurs. Try sweeping the telescope gently from side to side and it should appear as a very small out-of-focus star, perhaps slightly greenish-blue in color. At a magnitude of 10.4, it can be seen even in an 8 cm telescope if conditions are right.

Our last planetary nebula, **NGC 6026**, was once classified as a galaxy, but then, in 1955, the astronomer G. de Vaucouleurs correctly identified it. It is large, 40 arcseconds across, with a low magnitude, 12.5. An aperture of at least 20 cm will be needed to find it, and even then it will appear as a faint uniformly illuminated haze (see Figure 4.7). It lies midway between **Chi (χ) Lupi** and **Theta (θ) Lupi**.

There are several clusters in Lupus, both open and globular, although some of the open clusters are faint and small. Our first globular cluster is **NGC 5824**. This is a strongly condensed cluster, about 6.2 arcminutes across and at 9th magnitude, and can be seen as a hazy patch with a small telescope (see Figure 4.8). However, with an aperture of, say, 25 cm, it will begin to be resolved into stars. It is believed that the outer and therefore fainter members of the cluster are made even fainter by obscuring dust in our Milky Way.

A much brighter globular cluster is **NGC 5986**. It lies between a 5th magnitude star **h Lupi**, and a line of three 7th and 8th magnitude stars at the constellation's eastern side (see

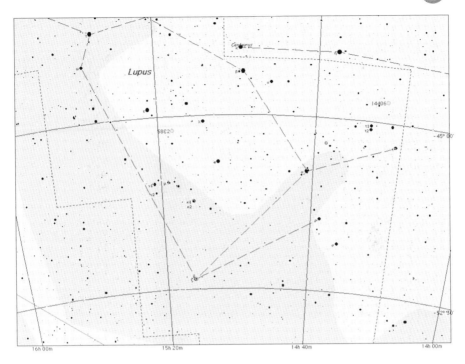

Star Chart 4.8. IC 4406; NGC 5882.

Figure 4.7. NGC 6026 (Space Telescope Science Institute, AAO, UK–PPARC, ROE, National Geographic Society, and California Institute of Technology).

Figure 4.8. NGC 5824 (Space Telescope Science Institute, AAO, UK–PPARC, ROE, National Geographic Society, and California Institute of Technology).

Star Chart 4.9). It is a small cluster at nearly 10 arcseconds diameter, and shines at 7th magnitude. In a small telescope it will appear as a large fuzzy disk of light, and is even visible in small binoculars. A 15 cm telescope at high magnification will readily resolve individual stars at the cluster's edge (see Figure 4.9).

A much more impressive cluster is **NGC 5927**. It lies about 3° northeast of **Zeta (ζ) Lupi** at the constellation's southeastern region (see Star Chart 4.10). Shining at a bright (for a globular) 8th magnitude and about 12 arcminutes in diameter it can be seen in even small telescopes, say 8 cm aperture, as a hazy spot. But with increasing aperture its central core, although unresolvable, becomes much brighter whilst the outer regions transform into a field full of sparkling stars (see Figure 4.10). It is a lovely object to observe. Current research suggests that the light from NGC 927 is dimmed by four magnitudes due to inter-stellar dust.

One of the few open clusters that we can look at with ease is **NGC 5822**. Due to its large size, some 40 arcminutes across,[5] it is best observed through binoculars or small telescopes in order to encompass all its component stars, which current estimates quote as 150 stars (see Star Chart 4.11). The total photographic magnitude of the cluster is 6.5, which means theoretically that it is on the threshold of naked-eye visibility. It would be interesting to know if this is so. Our final cluster is **NGC 5749**, which, for some obscure reason, is often omitted from many observing schedules. In truth it is not particularly bright, at 9th magnitude, or even large (only about 8 arcseconds across) and nor does it contain many stars – only about 35 – but it is a pleasing object nonetheless and in binoculars will appear as a small patch of dim haze, and in telescopes as a nice group of stars (see Figure 4.11). Observe it for yourselves and see what you make of it.

Now for something completely different – a dark nebula called **Barnard 228**. This is a long band of dark nebula, about 4° in length, easily spotted in binoculars. It lies halfway

---

[5] Larger than the full moon!

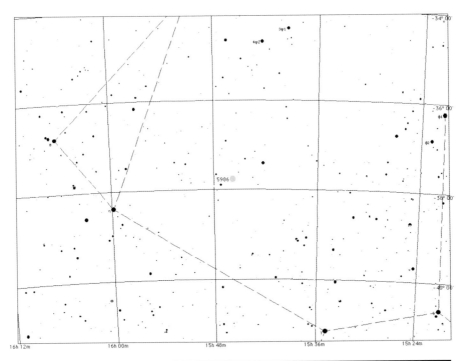

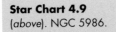

Star Chart 4.9
(*above*). NGC 5986.

Figure 4.9 (*right*).
NGC 5986 (Space
Telescope Science
Institute, AAO,
UK–PPARC, ROE,
National Geographic
Society, and California
Institute of Technology).

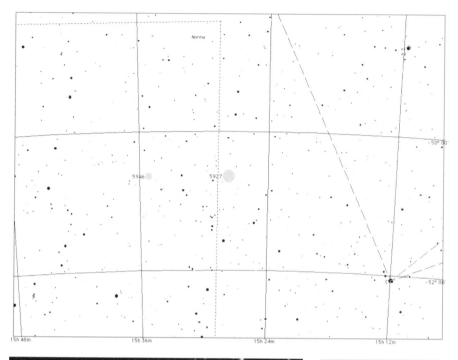

Star Chart 4.10
(*above*). NGC 5927.

Figure 4.10 (*left*).
NGC 5927 (Space
Telescope Science
Institute, AAO,
UK–PPARC, ROE,
National Geographic
Society, and California
Institute of Technology).

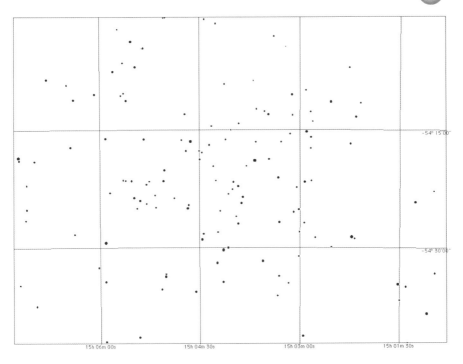

Star Chart 4.11. NGC 5822.

Figure 4.11. NGC 5749 (Space Telescope Science Institute, AAO, UK–PPARC, ROE, National Geographic Society, and California Institute of Technology).

between **Psi (ψ)** and **Chi (χ) Lupi.** It is best seen in low-power, large-aperture binoculars, as it will then stand out against the rich background star field.

4.4 Norma

Although this little constellation has the Milky Way running straight through it, as well as the galactic equator, and a great swathe of dark dust clouds, it is a little-known and fairly inconspicuous constellation (see Star Chart 4.12). Indeed, **Norma**, for that is its name, hasn't even got a star brighter than 4th magnitude, or even an Alpha or Beta Normae! So what has it got, you ask. Well, it does have a few double stars, some sparse open clusters, one bright globular cluster, and a few planetary nebulae, and of course the Milky Way! So with that in mind, let's start looking at some stars and the deep-sky objects.

A nice triple star, **Iota¹ (ι) Normae,** is our first object. The close pair will be a test in a 25–30 cm telescope, and were separated by about 0.5 arcseconds in 2000. The distant and fainter third member looks reddish by contrast to the brighter star. **Epsilon (ε) Normae,** on the other hand, is a nice wide, unequal magnitude – 4.8 and 7.5 – star. It has a lovely contrast in colors blue and yellow. Our third and final star is **h (Herschel) 4813.** This is a fine bright yellow star that seems to have a tiny white point some distance away. It will be seen easily in an 8 cm telescope and is set in a lovely star field which makes it even better.

The one globular cluster that we can see is **NGC 5946.** It is a small, fairly bright cluster and lies amidst a lovely star field that seems blessed with a number of wide pairs of stars

Star Chart 4.12. Norma.

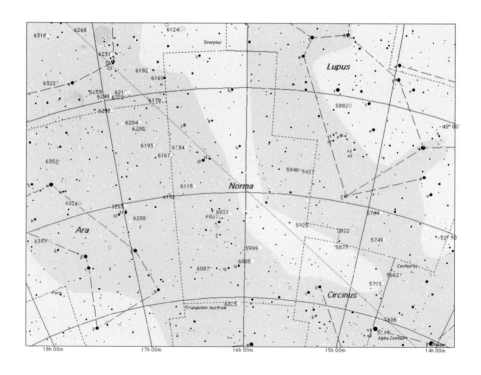

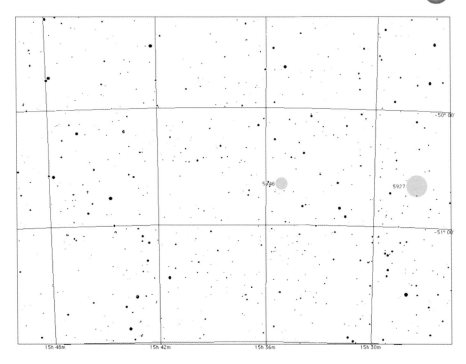

-50° 00'

-51° 00'

15h 48m   15h 42m   15h 36m   15h 30m

Star Chart 4.13. NGC 5946.

(see Star Chart 4.13). The cluster is just over 1 arcminute across and irregularly shaped. A 20 cm telescope will be needed in order to resolve the faint outer stars, whereas a 10 cm will just show it as a faint hazy blob (see Figure 4.12). Calculations suggest that the interstellar absorption here is nearly 2 magnitudes.

Several open star clusters are visible, but none of them present a spectacular view. Nevertheless we will still look at a few. **NGC 5999** is a fairly well spread out object with no bright stars (see Figure 4.13). It is about 10 arcminutes across and many of the stars seem to form lines and arcs.

On the other hand, **NGC 6067** is a group of about 100 stars that is very nice even in small apertures. Many of the stars are around 8th magnitude, along with a few fainter ones and most are packed into an area 14 arcseconds across. If you use a large aperture you may see a few more stragglers around its periphery, making the cluster about 20 arcseconds in diameter. In binoculars it will appear as a small round hazy glow set in a lovely star field (see Figure 4.14). This cluster deserves more attention than it gets.

A cluster that is also often forgotten is **NGC 6087**, which shines at about 5.5 magnitude. It is set again amidst a nice star field, with a bright yellow star at its center. There are about 40 stars making up the cluster of about 8th magnitude and fainter, all set in an area about 12 arcseconds across (see Figure 4.15). The bright star I mentioned above is in fact **S Normae**, a Cepheid variable star that varies in magnitude from 6.1 to 6.8 in about ten days.

Lying at Norma's eastern edge are two clusters, **NGC 6134** and **NGC 6152**. The former is a nice group of stars of 9th magnitude about 9 arcminutes across with a total magnitude of 7.2 (see Figure 4.16), whereas the latter is a lovely sight and perfect for binoculars. This is a group

Figure 4.12. NGC 5946 (Space Telescope Science Institute, AAO, UK–PPARC, ROE, National Geographic Society, and California Institute of Technology).

of around 70 or more stars shining at 8th magnitude and fainter and will appear as a lovely hazy glow, sprinkled through with just a few of the brighter members (see Figure 4.17).

A cluster that is very rarely mentioned is **Harvard 10**. It is a small group of around 35 stars, half of which can be seen in binoculars, with the rest contributing to the misty background (see Star Chart 4.14). Furthermore, located in and almost superimposed upon

Figure 4.13. NGC 5999 (Space Telescope Science Institute, AAO, UK–PPARC, ROE, National Geographic Society, and California Institute of Technology).

Figure 4.14. NGC 6067 (Space Telescope Science Institute, AAO, UK–PPARC, ROE, National Geographic Society, and California Institute of Technology).

Harvard 10's southwestern corner is **Collinder 299**, another cluster that will need something larger than binoculars. Indeed some amateurs are dubious as to whether a cluster is really there. What do you see? Another forgotten cluster is **Ruprecht 113**, which is somewhat difficult to pick out from the background stars. It is not very well defined as clusters go, and even though it is larger in diameter than the full moon, only boasts 15 stars of 8th and 9th magnitude. It makes one wonder –when is a group of stars an open cluster or an asterism?

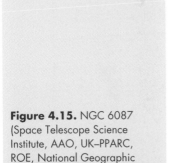

Figure 4.15. NGC 6087 (Space Telescope Science Institute, AAO, UK–PPARC, ROE, National Geographic Society, and California Institute of Technology).

Figure 4.16. NGC 6134 (Space Telescope Science Institute, AAO, UK–PPARC, ROE, National Geographic Society, and California Institute of Technology).

Finally there are two planetary nebulae we should look at. The easiest to observe is **Menzel 2 (Vorontsov–Velyaminov 78)**. It is, like so many other objects in Norma, set in a lovely star field. With a diameter of about 25 arcseconds, it can be seen in a telescope of about 15 cm as a grayish circular object. It lies some 1° southwest of the 5th magnitude star Kappa (κ) Normae (see Star Chart 4.15). It is roughly the same magnitude as Shapley 1 (see below), but because it is much smaller, it has a higher surface brightness and so is more easily observed. It will appear as a round, pale patch of gray in a nice star field, and has no central star.

Another fine planetary, and maybe the best in the constellation, is **Shapley 1**. It is one of the most elegant and beautiful nebulae in the entire sky, with a large diameter of about 80 arcseconds. This makes it similar in size to its famous northern cousin the Ring Nebula, M57, in Lyra. However it is much fainter, as its 13th magnitude light is spread out over a large area, and so it has a low surface brightness. It can be glimpsed with an 8 cm telescope

Figure 4.17. NGC 6152 (Space Telescope Science Institute, AAO, UK–PPARC, ROE, National Geographic Society, and California Institute of Technology).

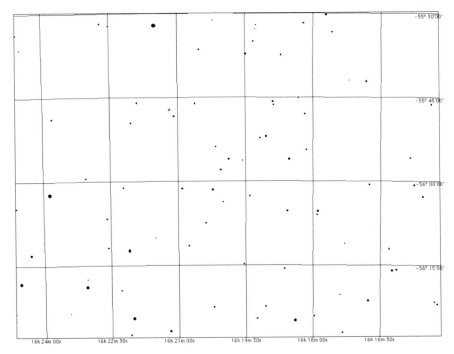

Star Chart **4.14** (*above*). Harvard 10.
Star Chart **4.15** (*below*). Menzel 2.

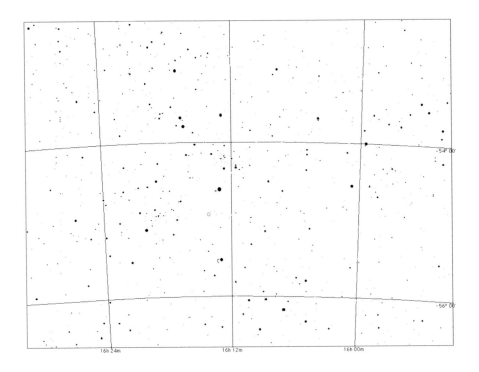

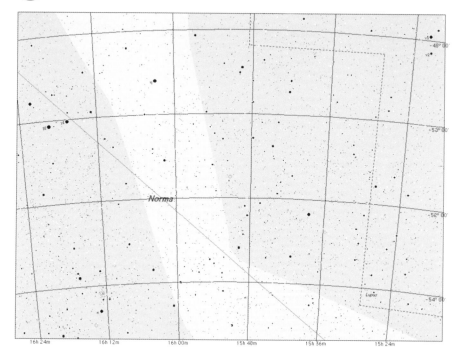

Star Chart 4.16. Shapley 1.

on very good dark evenings, otherwise only a 30 cm telescope will show the almost perfect ring shape (see Star Chart 4.16).

Our final object in Norma is the quite extraordinary bipolar nebula, **NGC 6164-65.** It surrounds the hot star **HD 148937**, which may or may not be a Wolf–Rayet star. It looks as if the nebula has been ejected as two elliptically shaped lobes from the southeast and northwest of the star. Each lobe is approximately 2 × 1 arcminutes. A 20 cm telescope will show its overall form, albeit faintly, and the use of a UHC filter will give a markedly visible improvement. This is a nice and unusual object with which to say a fond farewell to Norma and a hello to Ara.

4.5 Ara

This constellation is something of a conundrum. To the naked eye, **Ara** is a most inconspicuous area of the sky (see Star Chart 4.17). Its brightest stars are only of third magnitude so you might think that there is nothing worthwhile here to look at. You would be in error. The Milky Way is very rich here, with bright nebulosity, rich star fields and bands of dark nebula, and even a casual perusal with small binoculars will show glorious and rich vistas. In addition to the vast panoramas I have just mentioned, the constellation has many open and globular clusters, and, surprisingly for such a dense Milky Way region, several faint galaxies.

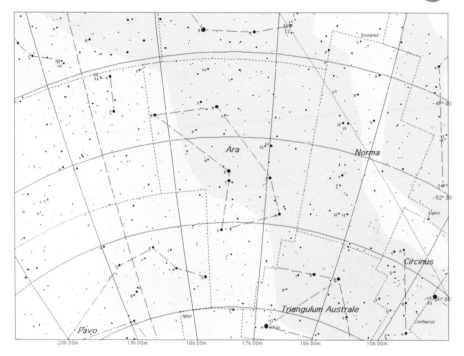

Star Chart 4.17. Ara.

In fact, an astute observer, and perhaps one who is already at home navigating about the sky, will know that we are now approaching some of the most spectacular parts of the Milky Way and that Ara is just a prelude for what is to come. The constellation transits in early June.

As is usual, we will begin with double stars, many of which show nice color contrasts. Our first that can be seen is the variable and double star **R Arae**. This consists of a nice yellow close pair, with a red star some 4 arcminutes north. The yellowish pair are separated by about 3.5 arcseconds, with magnitudes 6 and 8.5. There is some evidence that the primary star is also an eclipsing binary, so what we have here is a true triple star system. Set amidst a lovely star field is **CorO 201**, consisting of two pale yellow, equal magnitude stars that can be easily resolved in an 8 cm telescope. The system is a true binary star. A rather beautiful pair of stars is **Brs 13** which presents a deep golden-yellow and orange contrast. The stars are of magnitude 5.66 and 5.47 and can be easily resolved with apertures of about 8 cm. They are set against a nice but faint star-strewn field. A brilliant white star with a faint companion can be seen in **Gamma (γ) Arae**, a true binary system that under good conditions can just be resolved in a 10 cm telescope. An easy target for telescopes of 8 cm aperture is **h (Herschel) 4949**. This white pair of stars, magnitudes 5.7 and 6.44 are separated by about 2 arcseconds and are located within a nice star field.

Another fine color contrast can be seen in **I 40**. It consists of a bright yellow star with a somewhat fainter orange star some 4 arcminutes southeast. The yellow star itself has a companion located at its southwest that a 10 cm telescope will show under good conditions.

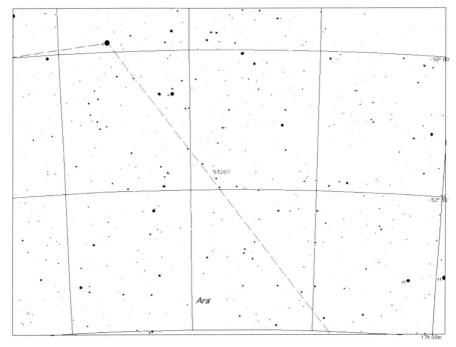

63260

Ara

-50° 00'

52° 00'

17h 00m

Star Chart 4.18. NGC 6326.

Our final double star is the fine well-separated pair **h** (**Herschel**) **4978**. One star has a nice pale yellow tint, the other is slightly off-white, with magnitudes 5.68 and 9.40, respectively. A small telescope of, say, 8 cm will resolve the system. Incidentally, the brighter of the two stars is a spectroscopic and eclipsing binary star, also known as **V539 Arae**, which has a period of 3.17 days.

There are a couple of planetary nebulae in Ara that we should look at. The first is **NGC 6326**. It is a small and circular object that cannot be missed, situated as it is in a nice star field. It is fairly well defined and gets perceptibly brighter towards the center. For those who need an aid to identification, there are two stars nearby located to its north and east (see Star Chart 4.18). It can be just glimpsed in a 20 cm telescope, but larger apertures will have no difficulty.

The second planetary is **Shapley 3**. This is a much fainter object and will need a 20 cm telescope at least for its circular shape to be appreciated. It will appear as a ghostly pale glow some 30 arcseconds in diameter (see Star Chart 4.19). A central star of 12th magnitude can be seen, providing conditions are right. A high power is recommended in order to see the central star.

Clusters are in no short supply here, and we will begin by looking at the globular clusters. Our first is **NGC 6352** (**Caldwell 81**), which is one of the less-condensed types of globular cluster. With a 15 cm telescope some of its stars are seen, and with a 20 cm telescope it will appear about 4 arcminutes in diameter with an irregular shape (see Figure 4.18). A slight but definite brightening toward its center will also become apparent. In even larger telescopes many more of the 12th magnitude members become resolved. It can actually be seen in binoculars as a small hazy glow set amidst the star field (see Star Chart 4.20).

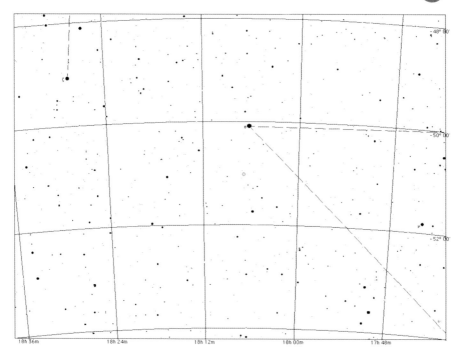

Star Chart 4.19 (*above*). Shapley 3.
Star Chart 4.20 (*below*). NGC 6352.

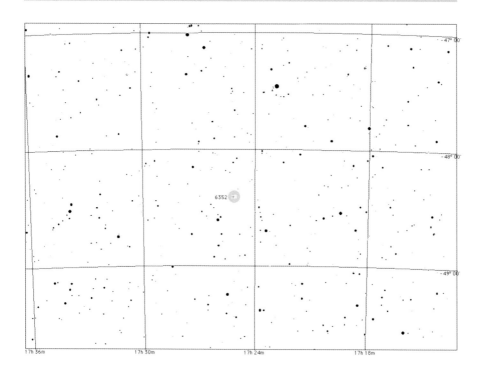

Figure 4.18. NGC 6352 (Space Telescope Science Institute, AAO, UK–PPARC, ROE, National Geographic Society, and California Institute of Technology).

Another fine globular cluster is **NGC 6362**. With a telescope of 10 cm aperture, the resolution will just be hinted as a granularity to the object, but with larger apertures the resolution is much more apparent as the cluster is resolved into many faint stars set on a hazy background (see Figure 4.19). The cluster is about 4 arcminutes in diameter and increases in brightness towards its center.

Figure 4.19. NGC 6362 (Space Telescope Science Institute, AAO, UK–PPARC, ROE, National Geographic Society, and California Institute of Technology).

Figure 4.20. NGC 6397 (Space Telescope Science Institute, AAO, UK–PPARC, ROE, National Geographic Society, and California Institute of Technology).

Now for something rather special. The globular cluster **NGC 6397** is a remarkable object to observe. Visible to the naked eye as a misty spot, it can be seen with binoculars and if larger binoculars are used a hint of resolution can be imagined (see Figure 4.20). It has a diameter comparable to that of the full moon, nearly half a degree, and with a telescope of, say, 8 cm some resolution is seen. It is an open cluster,[6] which incidentally helps us to see its individual stars, many of which are colored pale orange, and are displayed as arcs and chains. It is probably the second-nearest globular cluster to us, with only Messier 4 in Scorpius being nearer. If this cluster were in the northern part of the sky, it would be much more well known than it is today. Alas it is not, and remains an object of delight for southern observers (see Star Chart 4.21).

Many splendid open clusters are also located with the boundaries of Ara. A fine example for small telescopes is **NGC 6200**. It is around 10 arcminutes in size and has two bright stars to the north of its center whilst its component stars are arranged in arcs and chains (see Figure 4.21). As an aside, a little irregular group of stars directed towards a yellow star of 7th magnitude is a separate cluster known as **Hogg 20**.

An even more unassuming group of stars some 10 arcminutes from the star is known as **Hogg 21**. I find it difficult, if nigh on impossible, to distinguish these small clusters from the background's star fields. What do you see?

A small but bright cluster is **NGC 6204** and it is immediately apparent as it stands out from the background field (see Figure 4.22). It is, however, only 5 arcminutes across and fairly scattered. A small group of stars some 7 arcminutes to the southeast are known as **Hogg 22**. Incidentally, it is often possible to get both NGC 6204 and NGC 6200 in the same field of view.

A small cluster that is perfect for binoculars is **NGC 6208**. It is located to the north of **Zeta (ζ) Arae** and will appear as a 7th magnitude fuzzy haze formed of about 70 stars (see Star Chart 4.22). A very easy cluster to observe is **IC 4651** (see Figure 4.23). This is an

---

[6] This doesn't mean the same as a galactic cluster, but rather that the stars are well spread out.

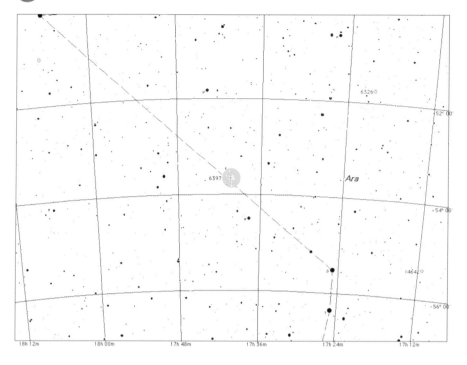

Star Chart 4.21. NGC 6397.

Figure 4.21. NGC 6200 (Space Telescope Science Institute, AAO, UK–PPARC, ROE, National Geographic Society, and California Institute of Technology).

Figure 4.22. NGC 6204 (Space Telescope Science Institute, AAO, UK–PPARC, ROE, National Geographic Society, and California Institute of Technology).

irregular group of stars which are spread through a region of 15 arcminutes, although a few fainter members reach out to 20 arcminutes. It can be glimpsed in binoculars as a hazy spot of about 7th magnitude, and looks even better in small to medium-aperture telescopes (see Star Chart 4.22).

A cluster that has a nice star group in its center is **NGC 6193**. The cluster itself has about 35 members, many of which can be seen through binoculars (see Figure 4.24). The stars range in magnitude from 5th to 11th and can be seen in moderate-aperture telescopes. The group is called **h 4876**. What makes this cluster special, however, is that it is part of what is known as the **NGC 6188–93 Complex**. This is a large expanse of sky comprising both the cluster NGC 6193 and nebulosity **NGC 6188**.

The cluster itself is rather large and will need a low power in order to be fully encompassed at the eyepiece. It consists of many loops and arcs of stars and indeed is a perfect example of a showpiece cluster, even in small telescopes. However, there is a fair amount of nebulosity associated with the cluster spanning more than 3°. It consists of both dark dust clouds and bright emission nebulae, the brightest part of which is given the NGC catalogue number. It is, as expected, very difficult, but not impossible, to observe. It needs a dark and transparent night, a large telescope, and perhaps an [OIII] filter, in order for it to be glimpsed. Try it and let me know what you see.

We will finish our tour of Ara by looking at two galaxies. Bear in mind that to see galaxies at such low galactic latitude is rare, as we would expect the light coming from them to be extinguished by the dust and gas of the Milky Way. This would imply that interstellar absorption in this area of the Milky Way is relatively low.

The two galaxies are **NGC 6221** and **NGC 6215**. The former is the brighter of the two and is close to **Eta (η) Arae**, about 22.5 arcminutes southeast of the orange star. A 30 cm telescope will show a small object some 2.5 × 1.5 arcminutes in size with a bright center (see Figure 4.25). Using a high power some evidence of a spiral arm may be glimpsed.

Located only 10 arcminutes west from Eta Arae is the fainter galaxy, **NGC 6215**. This will appear as a small round object only 1 arcminute across with a bright center (see Figure

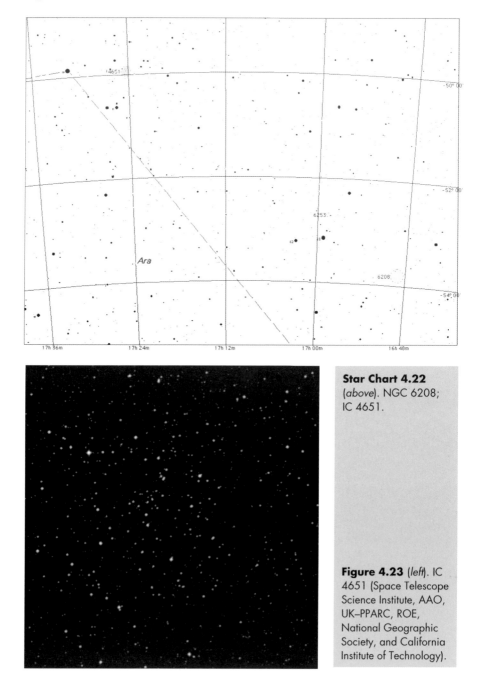

Star Chart 4.22
(*above*). NGC 6208;
IC 4651.

Figure 4.23 (*left*). IC
4651 (Space Telescope
Science Institute, AAO,
UK–PPARC, ROE,
National Geographic
Society, and California
Institute of Technology).

4.26). Both of these galaxies can be seen in a 15 cm telescope. It appears that they are members of a small group of galaxies about 15 million parsecs away. Incidentally, both galaxies lie in the same field as Eta Arae (see Star Chart 4.23).

Figure 4.24. NGC 6193 (Space Telescope Science Institute, AAO, UK–PPARC, ROE, National Geographic Society, and California Institute of Technology).

4.6 Pavo

You may think that the constellation **Pavo** should actually be in Book 1 as it transits in mid-July, and so it does. However, the Milky Way only runs through its westernmost

Figure 4.25. NGC 6221 (Space Telescope Science Institute, AAO, UK–PPARC, ROE, National Geographic Society, and California Institute of Technology).

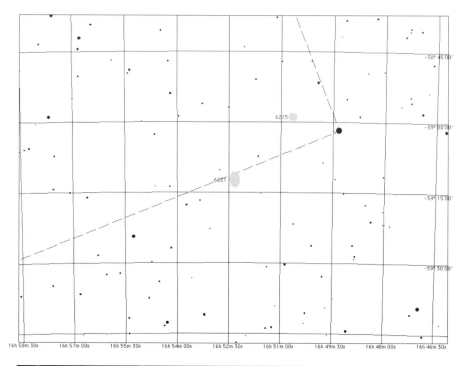

16h 58m 30s   16h 57m 00s   16h 55m 30s   16h 54m 00s   16h 52m 30s   16h 51m 00s   16h 49m 30s   16h 48m 00s   16h 46m 30s

Star Chart 4.23
(*above*). NGC 6221;
NGC 6215.

Figure 4.26 (*left*).
NGC 6215 (Space
Telescope Science
Institute, AAO,
UK–PPARC, ROE,
National Geographic
Society, and California
Institute of Technology).

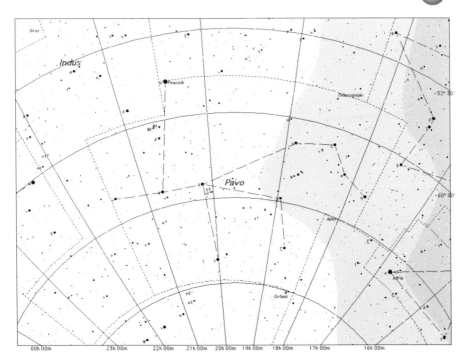

Star Chart 4.24. Pavo.

region, and so this part of the constellation transits in June. Thus, I place it in this chapter (see Star Chart 4.24).

Let's start with a few stars: two variable, a double and a very red-colored star. The first is **Lambda (λ) Pavonis,**[7] one of the brightest irregular variable stars in the entire sky. It varies, erratically, in magnitude from 3.4 to 4.4 and is classified as a **Gamma Cassiopeiae** type variable. Because it so bright, its light variations can be observed with the naked eye. The second star is **Kappa (κ) Pavonis.** This is a class of Cepheid variable star that is further subclassed as a **W Virginis,** or Type II class. It has a period of 9 days and varies from 4th to 5th magnitude.

The double star is **Xi (ξ) Pavonis.** This consists of a bright orange star, magnitude 4.5, with a nice white companion. Both are set in a lovely star field and can be resolved with an 8 cm telescope. The bright star is itself a spectroscopic binary.

There are several intensely red stars in the sky, and one of them is **V Pavonis.** This is a red giant variable star, of the class SRB, that varies in magnitude from 6.3 to 8.2 over a period of 225.4 days. It also has a secondary period of about 3735 days. However, what makes the star so special is that it is a glorious deep red color and really is quite spectacular. Often, many amateurs are disappointed after reading about a star's color and then seeing it for themselves, but this is one star where you will be very pleased. Observe this and enjoy it.

---

[7] Actually, the star lies on the exact border of the Milky Way, so I have decided to mention it here.

Figure 4.27. IC 4662 (Space Telescope Science Institute, AAO, UK–PPARC, ROE, National Geographic Society, and California Institute of Technology).

Pavo also has quite a few galaxies within its borders, and surprisingly some of these actually lie within the confines of the Milky Way. These are the ones we shall concern ourselves with here. Close to the star **Eta (η) Pavonis** is a small dwarf galaxy **IC 4662**. It is only 1 arcminute in size and so will require a high magnification in order to see any structure (see Figure 4.27). Some reports indicate that with good conditions, and using either an [OIII] or UHC filter, it is possible to see an HII region on the eastern edge of the galaxy. It is situated 10 arcminutes northeast of the aforementioned star.

A test for 30 cm telescopes is **NGC 6630**. You will need a dark and transparent night in order to observe this elusive object. It will appear as a very faint, circular haze about 20 arcseconds across (see Star Chart 4.25).

Lying very close to it about 20 arcminutes southeast is another faint galaxy, **IC 4723**. Incidentally, both these galaxies were initially believed to be planetary nebulae, and in older catalogues were referenced as such (see Star Chart 4.25).

Our final object is the galaxy **NGC 6684** which is a nice round and symmetrical type Sa galaxy (see Figure 4.28). It is only 45 arcseconds across, however, and so you will need care to locate and identify it. A nearby star, **Theta (θ) Pavonis**, may be a mixed blessing as it does act as a pointer to the galaxy, but its glare may prevent identification (see Star Chart 4.26). A 15 cm telescope should be able to deal with this faint object.

4.7 Libra

For completeness, we must mention the constellation **Libra** (see Star Chart 4.27). Most books state that the constellation lies just outside the Milky Way, but in reality its eastern regions do reside within the Milky Way. However, that means that there are few objects that concern us. There are no open clusters or emission nebulae and only a handful of

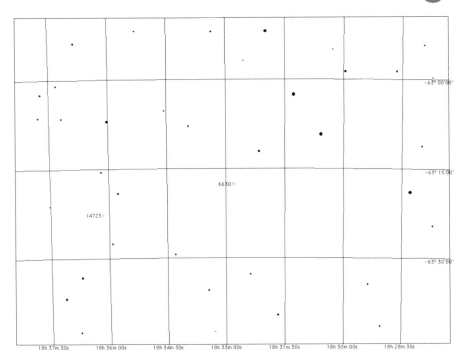

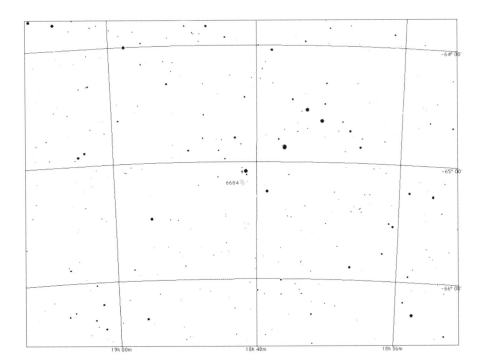

Star Chart 4.25 (*above*). NGC 6630; IC 4723.
Star Chart 4.26 (*below*). NGC 6684.

Figure 4.28. NGC 6684 (Space Telescope Science Institute, AAO, UK–PPARC, ROE, National Geographic Society, and California Institute of Technology).

galaxies. Strangely enough, deep imaging has revealed that the constellation is actually full of galactic reflection nebulosity, but this, for the most part, is beyond amateur equipment.[8]

There is a globular cluster, **NGC 5897**, which is large and loosely structured. Its magnitude is about 8.6, but because it is spread out over nearly a quarter of a degree, its surface brightness is low, thus making it difficult to observe (see Figure 4.29). A really dark sky will be needed for a 10 cm telescope and even with a 15 cm aperture all that will be seen is a dim, patch of gray light. Larger apertures will show the cluster's definite circular shape. There is some debate as to whether small telescopes can resolve the stars within NGC 5897. Some reports state that a 10 cm will just resolve stars, and others say that a larger telescope is needed. It has to be said that this is not one of the best globulars to observe in the sky, but the challenge for you here is just to be able to locate it (see Star Chart 4.28).

Located several degrees to the southeast of NGC 5897 is the planetary nebula **Me (Merrill) 2–1 (PK 342 +27.1)**. This was discovered in the 1940s by the astronomer Paul Merrill. It is about 7 arcseconds in diameter, and has a bluish disk, but small telescopes will only show it as a star, whereas larger apertures, along with a high magnification (maybe as high as 300×) will reveal its true origins. There is a central star but it is a very faint 15.4 magnitude. The best way to find this elusive object is to use a low to medium power and find its approximate position (see Star Chart 4.29). Then see if you can locate something that is just slightly out of focus – this should be the planetary.

A few galaxies are visible to us, but again they will need telescopes of at least 20 cm in order to be glimpsed. The first is **NGC 5898**, which has a round shape, just over 1 arcminute in size with a faint but discernable brightening at its center. Some

---

[8] No doubt this is being imaged with an amateur telescope as I write this!

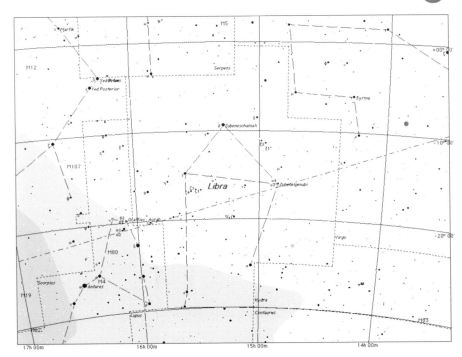

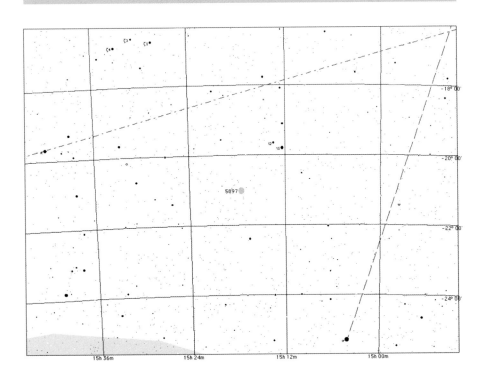

Star Chart 4.27 (*above*). Libra.
Star Chart 4.28 (*below*). NGC 5897.

Figure 4.29. NGC 5897 (Space Telescope Science Institute, AAO, UK–PPARC, ROE, National Geographic Society, and California Institute of Technology).

Star Chart 4.29. Merrill 2–1.

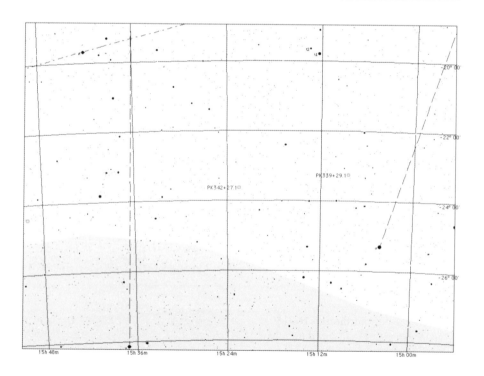

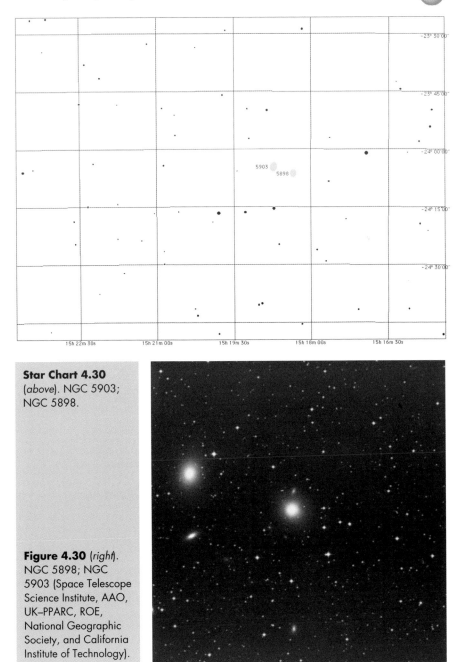

Star Chart 4.30 (*above*). NGC 5903; NGC 5898.

Figure 4.30 (*right*). NGC 5898; NGC 5903 (Space Telescope Science Institute, AAO, UK–PPARC, ROE, National Geographic Society, and California Institute of Technology).

5 arcminutes to its northeast is **NGC 5903**. It has a brightness similar to NGC 5898, but has a more elongated appearance some 1.25 × 0.75 arcminutes along with a stellar-like nucleus (see Star Chart 4.30 and Figure 4.30).

4.8 Scorpius

Let's now look at one of the most spectacular parts of the Milky Way, the constellation **Scorpius** (see Star Chart 4.31). This is one of the few constellations that actually resemble the object it represents – a scorpion. Although it is considered a southern constellation, it does rise high enough into the sky so that northern observers can at least get a glimpse of many of the treasures it contains (see Figure 4.31). It transits in early June. The constellation also lies in the general direction of the center of our Galaxy (see Sagittarius).

It is not an exaggeration for me to say that quite a considerable part of this book could be devoted just to this one part of the sky. It contains a lot of objects that warrant observation, ranging from naked-eye globular clusters to nebulae that can only be glimpsed with the largest telescopes. It is perfect for observing on cool summer evenings. Try to spend several nights just scanning the area with binoculars and a telescope to get a feel for the magnificence of this area.

Many of the objects here need a dark sky, which can be said for most astronomical targets, but more so when the Milky Way lies close to the horizon. Try to find a very dark site and you will be amply rewarded.

There are several superb double stars in Scorpius, so let's start with perhaps the most famous in the constellation, **Antares**, or **Alpha (α) Scorpii**. A gloriously colored star of fiery red (or, as some astronomers of the last century observed, saffron-rose), it contrasts nicely with its fainter green companion. The sixteenth-brightest star in the sky, this is a red giant, with a luminosity 6000 times that of the Sun, and a diameter hundreds of times

Star Chart 4.31. Scorpius.

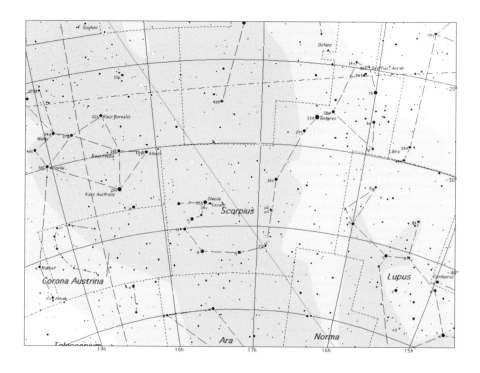

Figure 4.31. The Milky Way in Scorpius (Matt BenDaniel, http://starmatt.com).

bigger than the Sun's. But what makes this star especially worthy is the vivid color contrast that is seen between it and its companion star. It can be glimpsed with an aperture of 8 cm, but more often than not the bright glare from Antares often inhibits observation. Nevertheless it is a wonderful star. The companion has a magnitude of 5.4, lying 2.6 arc-seconds away. Antares is also the brightest member of a group of about 20 stars, all in the same general area, called the **Antares Cluster**. They all exhibit the same proper motion and are part of the much larger **Scorpius–Centaurus OB Association**.

We are actually located at the interior edge of the **Orion–Cygnus Spiral Arm** and the Scorpius–Centaurus OB Association can represent the very inner edge of this arm. So when we look through the Scorpius–Centaurus OB Association we are in fact looking across an interarm gap toward the next spiral arm of our Galaxy, which is known as the **Sagittarius–Carina Spiral Arm.**

Another fine double star is **Beta (β) Scorpii**. This is a good double for small apertures of around 5.0 cm and has a brilliant blue-white primary with a paler blue secondary. The primary is itself a spectroscopic binary, so in fact we are looking at a quadruple system. The magnitudes of the stars are 2.6 and 4.9, separated by 13.6 arcseconds. An interesting star is **Nu (ν) Scorpii**, which is a system of double-double stars. The main white pair will appear as two disks in contact, but only under near perfect conditions; whilst the easier pair are a nice pale yellowish color. The primary should be resolved in a 15 cm telescope, whilst the secondary can be glimpsed in an 8 cm telescope. The two components appear as magnitude 4.2 and 6.1, with a separation of 41.1 arcseconds.

There are several fine stars to observe, no matter what equipment you use. One of these is **2 Scorpii**. This bright star, magnitude 4.69, has a close secondary about 3 arcseconds to its west at magnitude 7.12. Another is **12 Scorpii** separated by 4 arcminutes, magnitudes 5.79 and 8.13. **HN 39** is another fine double set amidst a lovely star field separated by about 6.5 arcseconds.

A star that may prove a challenge is **Sigma (σ) Scorpii**. The lovely bright white star has a faint companion to its west at around magnitude 5. It is exceedingly close, perhaps only half an arcsecond, so will need apertures around 30 cm for resolution. A faint nebula also envelops the star, but this is also difficult to detect and may need the use of an Hβ filter.

A lovely double star situated at the end of the cluster **Trumpler 24** is **I 576**. This is a bright white star that contrasts superbly with the background star field. A star that has a nice color contrast is **Howe 86.** A bright yellow primary along with an orange-red secondary, it is easy to see in an 8 cm telescope. Another orange-red star that is not only set in a star-sprinkled field but has two faint white companions is **h (Herschel) 4926.** The primary star here is a red supergiant that has its light reduced by about two magnitudes due to the dust in the Milky Way. The two secondary stars can be seen easily in a 10 cm telescope, but only the slightly brighter one will be seen in anything smaller.

An easy naked-eye double star is **Zeta^1** and **Zeta ^{2} (ζ) Scorpii. Zeta^1** is a blue supergiant, magnitude 4.7, and **Zeta^2** is an orange star with a magnitude of 3.62. Some observers have called into question the color of Zeta^1 when seen with the naked eye or a small telescope saying it looks more yellow than blue. Try for yourself and see what color you think it is. In addition, it is possible that Zeta^1 is in fact an outlying member of the cluster NGC 6231. Another nice naked-eye double is **Mu^2** and **Mu^2 (μ) Scorpii** consisting of two blue-white stars, magnitudes 3 and 3.56. **Mu^1** is in fact an eclipsing binary star with a magnitude range of only 0.3.

Our final double star is **Xi (ξ) Scorpii**. To call it a double is a misnomer, as it is an excellent quadruple star for small telescopes. In a telescope it will appear as a 4th magnitude star with a 7th magnitude companion about 7 arcseconds away. But also visible is the same field is a much wider pair of stars of 7th and 8th magnitude, known as **Σ (Struve) 1999**. It appears that all four stars are gravitationally bound.

There are several variable stars in Scorpius easily within the range of amateur equipment. One of these is the splendid **RR Scorpii**, considered by many to be the finest in the constellation. It is a long-period variable, class M,[9] that ranges in magnitude from a naked-eye 5.11[10] to a faint 12.3, over a period of some 279 days. Another long-period variable is **BM Scorpii**, a member of the type of variable class known as a semiregular yellow supergiant. This has a nice orange tint, and its spectral class varies during its periods of variability from K0 to K3. The magnitude range is 6.0 to 8.1 over a period of some 850 days. Both of these stars are pulsating variables, which means that they actually alter in size, which is responsible for the change in brightness, or luminosity.

Scorpius is very lucky in having many fine clusters, both open and globular. In fact, there are over 60 of them! To describe all of them would be an onerous task, so I shall just pick those that are the brightest and most interesting to us. That is not to say that the ones left out are not worth bothering about – far from it. All the clusters that lie herein are worth observing, but time and space make the decision of what to observe. Let us begin.

Our first port of call is **NGC 6231 (Caldwell 76)**, a superb cluster located in an awe-inspiring region of the sky. This is a very large open cluster about 20 arcminutes across

---

[9] So called because it is similar to the archetypal star of its class, Mira.
[10] Some sources quote this to be 6.97, but the observations may have been made as the star was beginning to fade.

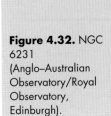

Figure 4.32. NGC 6231 (Anglo–Australian Observatory/Royal Observatory, Edinburgh).

that contains many fine stars. There are some 130 stars within the cluster and although only around 15–20 will actually be visible – those that are between 5th and 6th magnitude – it is an easy target for binoculars (see Figure 4.32). It is brighter by 2.5 magnitudes than its northern cousins, the double cluster in Perseus. The cluster is full of spectacular stars: very hot and luminous O-type and B0-type giants and supergiants, a couple of Wolf–Rayet stars, and ξ^{-1} Scorpii, which is a B1.5 Ia extreme supergiant star with a luminosity nearly 280,000 times that of the Sun! One star in particular is **van den Bos 1833 (B1833)**, which is a nice double star of magnitudes 5.6 and 7.3. The cluster is thought to be a member of the stellar association Sco OB1, with an estimated age of 3 million years. It can be seen with the naked eye as a hazy spot and is a wonderful object in binoculars and telescopes, the cluster containing many blue, orange and yellow stars. It lies between **Mu^{1+2} (μ^{1+2}) Scorpii** and **Zeta^1 (ξ^{-1}) Scorpii**, an area rich in spectacular views. This is a good cluster to test the technique of averted vision, where many more stars will jump into view. Just looking at this region alone is worthwhile as it is strewn with colored gems and dazzling star fields. Try to spend some time just looking at this lovely part of the Milky Way, as we are looking at what is a Sagittarius–Carina Spiral Arm tracer (see earlier entry on Antares).

Another fine cluster is **NGC 6124 (Caldwell 75)**, suitable for large binoculars and small telescopes and spread over 25 arcminutes, nearly the size of the full moon. It will need a large field of view to be seen in its entirety and there is a chain of stars at its southern edge, and a tightly grouped collection of five bright stars at its center (see Figure 4.33). It also contains several other nice star chains and a few red-tinted stars. There are about 30–40 stars that can be seen in large binoculars, when many of its 9th magnitude stars will be glimpsed. If a large-aperture telescope is used, say 30 cm, then over 100 stars can be seen. It is relatively close, at a distance of around 1500 light years.

Lying between **Mu (μ) Scorpii** and Collinder 316 is a scattered group of stars known as **NGC 6242**, some 10 arcminutes across. With binoculars it will appear as a single 6th magnitude star with four or five fainter 9th magnitude attendants set against the almost resolvable but much fainter haze of the other members of the cluster. In a small telescope of 10 cm aperture it is a nice object.

A superb and often quoted as a beautiful cluster is **NGC 6281**. In binoculars there are about seven stars set in a shape resembling a cross, of magnitudes 7–9. But in a small telescope of around 10 cm, there is a nice pyramid-shaped feature about 8 arcminutes across at the cluster's center (see Figure 4.34). There are also some nice pairs of stars and two orange stars in particular. A much smaller and fainter cluster is **NGC 6322**. This lies between **Theta (θ) Scorpii** and **Eta (η) Scorpii**. In binoculars it will only be seen as a group

Figure 4.33. NGC 6124 (Space Telescope Science Institute, AAO, UK–PPARC, ROE, National Geographic Society, and California Institute of Technology).

of about five or six stars set against the haze of the remaining fainter stars, but it is a pretty sight in small telescopes where more of its 30 members will be seen.

Easily seen with the naked eye as a dim patch of light is the cluster **Messier 6 (NGC 6405)**, also known as the **Butterfly Cluster**. It is, in my opinion, one of the few stellar objects that actually looks like the entity after which it is named. A fine sight in binoculars, it contains the lovely orange-tinted star BM Scorpii east of its center (see earlier). Small binoculars will reveal about 30 of its stars in a field of view some 15 arcminutes across, whilst larger binoculars and small telescopes will reveal more of its 80 members (see

Figure 4.34. NGC 6281 (Space Telescope Science Institute, AAO, UK–PPARC, ROE, National Geographic Society, and California Institute of Technology).

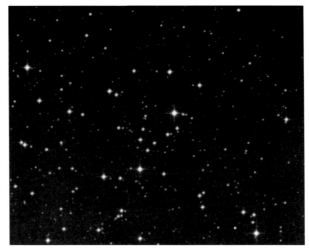

Figure 4.35. Messier 6 (Space Telescope Science Institute, AAO, UK–PPARC, ROE, National Geographic Society, and California Institute of Technology).

Figure 4.35). It is spread over about 25 arcminutes, so a large field of view is needed to really appreciate this fine cluster. Surrounding the cluster are many nice steely blue-white stars. It is believed to be at a distance of 1590 light years.

A small and often passed-over open cluster is **NGC 6451**, which admittedly is at times difficult to pick out from the stunning star fields of the Milky Way. It is a group about 6 arcminutes across with a perceptible dark rift across it from north to south (see Figure 4.36). It isn't bright, and you will need a 15 cm aperture telescope to appreciate it, but it is a nice object nonetheless.

Figure 4.36. NGC 6451 (Space Telescope Science Institute, AAO, UK–PPARC, ROE, National Geographic Society, and California Institute of Technology).

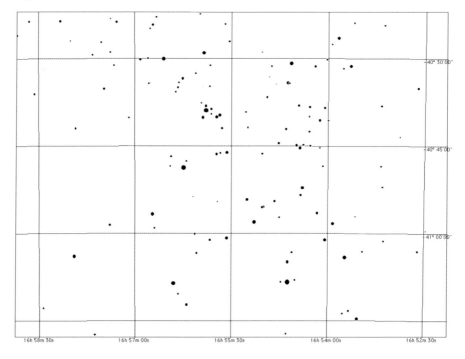

Star Chart 4.32. Collinder 316.

Our penultimate clusters are the small and little-known **Collinder 316** and **Trumpler 24**. The cluster Collinder 316 is a large cluster that spreads over nearly 2 degrees, which is four times as large as the full moon (see Star Chart 4.32). Most of its stars range in magnitude from 6th to 9th and it is best appreciated, perhaps some would say only, with binoculars. A loose and scattered cluster, **Trumpler 24**, also known as **Harvard 12**, is set against the backdrop of the Milky Way, but is only half the size of Collinder 316. It contain around 200 stars and is associated with the faint nebulosity IC 4628 (see later). Both of these clusters are believed to be located at the core of the Scorpius OB1 stellar association.

Our final open cluster is the brightest and easily the most spectacular: **Messier 7 (NGC 6475)**. This is an enormous and spectacular cluster and presents a fine spectacle in binoculars and telescopes, containing over 80 blue-white and pale yellow stars. It is easily visible to the naked eye as a hazy glow some 4° northeast of **Lambda (λ) Scorpii**. In binoculars of any size it is a beautiful object and in small telescopes it appears as over 40 arcminutes across with many outlying stars (see Figure 4.37). It can be seen even in a finder scope and will need a wide field to be really appreciated. There is a fine orange star southwest of the cluster's center and many of the fainter stars have a pale yellow tint. It is only just over 800 light years away, but is over 200 million years old. Many of the stars are around 6th and 7th magnitude, and thus should be resolvable with the naked eye. Try it and see. Incidentally, located in the same field of view as M7 is the faint and distant globular cluster NGC 6453, an 11th magnitude hazy spot only 1 arcminute across.

Globular clusters in plenty are also found in this part of the Milky Way and our first is the cluster **Messier 80 (NGC 6093)**. Readily detectable in binoculars as a tiny 9 arcminute, glowing hazy patch set in a stunning star field, it has a distinctly noticeable brighter core

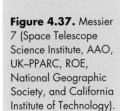

Figure 4.37. Messier 7 (Space Telescope Science Institute, AAO, UK–PPARC, ROE, National Geographic Society, and California Institute of Technology).

(see Figure 4.38). However, telescopes with an aperture of about 20 cm will be needed to resolve its 14th magnitude stellar core, while a 10 cm telescope will show it as a hazy circular spot. Along with its outlying stars it covers an area about 3 arcminutes across. It is one of the few globulars to have been the origin of a nova, **T Scorpii**, when it flared to prominence in 1860, then disappeared back into obscurity within three months.

A rather beautiful globular is **Messier 4 (NGC 6121)**. This is a superb object, presenting a spectacle in all optical instruments and even visible to the naked eye. But it does lie very close to the star Antares, only 1.5° away, so that the glare of the latter may prove a problem in detection (see Star Chart 4.33). Many of the outlying stars form gentle arcs that reach some 12 arcseconds across, which have reportedly been visible in telescopes as small as 8 cm aperture. High-power binoculars, however, will resolve several stars, providing conditions are ideal (see Figure 4.39). But telescopes of all apertures show detail and structure within the cluster, and the use of high magnification will prove beneficial; but what is more noticeable is the bright lane of stars that runs through the cluster's center. It is thought to be the closest globular to the Earth at 6500 light years (although NGC 6397 in Ara may be closer), and about 10 billion years old.

Figure 4.38. Messier 80 (Harald Strauss, AAS Gahberg).

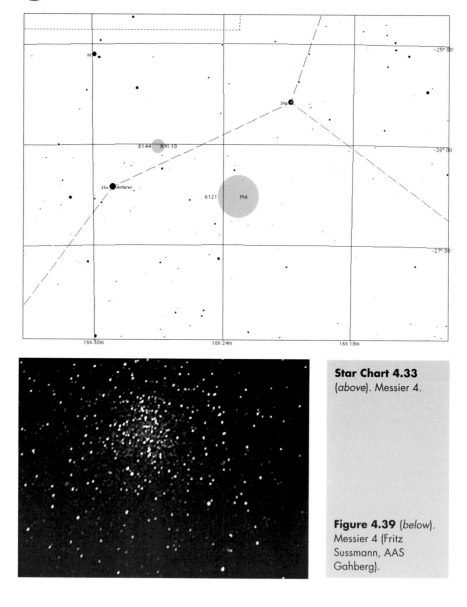

Figure 4.39 (*below*). Messier 4 (Fritz Sussmann, AAS Gahberg).

Two nice but faint globular clusters are **NGC 6144** and **NGC 6139.** The former is an irregular cluster that shows little central condensation and lies behind the reflection nebula that surrounds Antares. It is about 2 arcminutes across. A 15 cm telescope will show just a faint haze, along with a bright foreground star on its western edge, whilst a 20 cm telescope will just begin resolution (see Figure 4.40). The latter globular is rather more compressed and somewhat brighter, even though it is estimated that about 2.3 magnitudes of absorption due to dust is taking place. It is only 1.5 arcminutes across and with a 10 cm telescope will appear as a circular hazy spot (see Figure 4.41). Resolution of the globular becomes apparent with apertures of about 30 cm.

Figure 4.40. NGC 6144 (SBAS).

A difficult globular cluster to observe is **NGC 6256**. It is only 1 arcminute across and its stars are at individual magnitudes of around 15. Couple this with the cluster being termed sparse, and you see what I mean. It is visible in a 15 cm telescope, but a 30 cm aperture is need for resolution.

Another challenge for globular cluster seekers is the faint object **NGC 6380**. The cluster is near an 8.5 magnitude star that tends to impede observation. It is also small at 1 arcsecond across, and even though it is barely visible with a 15 cm telescope, there is no certainty that you'll see it with a 20 cm aperture (see Star Chart 4.34). Located in the same field as NGC 6380 is the faint and reddened globular **Ton 2**. It lies about 6 arcminutes to the north-

Figure 4.41. NGC 6139 (Space Telescope Science Institute, AAO, UK–PPARC, ROE, National Geographic Society, and California Institute of Technology).

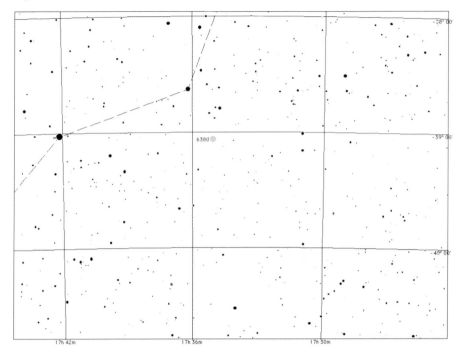

Star Chart 4.34. NGC 6380.

west of the bright star **Q Scorpii** and has a very low surface brightness. It can be seen with a 25 cm telescope, but larger apertures will find it more easily.

A globular cluster that can be seen in telescopes as small as 8 cm is **NGC 6388**. This is a bright and round, well-compressed object that has a mottled appearance. It is around 3 arcminutes across and although easy to see it needs a large aperture in order to be resolved (see Figure 4.42).

Another small-telescope cluster is **NGC 6441**. This is 2 arcminutes across and has a fine orange star, **G Scorpii**, about 4.5 arcminutes to its west. It will appear as a hazy glow that hints at resolution.

Located just west of M7 is the globular **NGC 6453**, a small cluster irregular in shape that will appear as a hazy spot. It is only 1 arcsecond across and will not appear resolved. Our final globular cluster is **NGC 6496**, which is a very open type cluster. Some stars can be seen but are believed to be field stars, and even a 30 cm telescope will only show a circular haze about 2 arcminutes in diameter. It can be glimpsed as a small hazy spot in a 10 cm telescope.

Scorpius has its fair share of planetary nebulae as well as clusters. Many of them are in reality quite faint and small, so only the brightest and largest will be mentioned here. Let's start with **NGC 6072**. Set in a nice star field, this is a bright but small nebula, only 40 arcseconds across. It is circular in shape, and shows a somewhat slight brightening towards its center (see Figure 4.43). It should be easy to see under a dark sky with a 20 cm telescope.

Another nice nebula is **NGC 6153**, which is about 20 arcseconds across, and shows a definite bluish-green color. It is quite easy to locate as it lies at the southern corner of a small diamond-shaped asterism of an orange star, a white star and a close double star. It should be easy to find even in an 8 cm telescope.

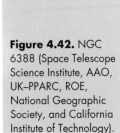

Figure 4.42. NGC 6388 (Space Telescope Science Institute, AAO, UK–PPARC, ROE, National Geographic Society, and California Institute of Technology).

A most remarkable planetary nebula is **NGC 6302**, also known as the **Bug Nebula**. It is a bluish colored nebula, elliptical in shape, some 1.5 to 0.5 arcminutes (see Star Chart 4.35). It is visible in nearly all sizes of telescope, and with a large aperture a distinct brightening at the center will be seen, which many incorrectly assume is the central star. In fact it is just the bright central region. A dust lane runs across the planetary nebula and obscures the central star from view (see Figure 4.44). It is a class of nebula called bipolar, which to some will resemble an extended butterfly. The star responsible for the nebula has a temperature of 380,000 K. Unfortunately, the nebula is not visible from the UK.

Another nice nebula is **NGC 6337**, which was described by John Herschel as "a beautiful delicate ring of a faint ghost-like appearance" (see Figure 4.45). It needs a large-aperture telescope in order to be seen, say a 20 cm or bigger (see Star Chart 4.36). A planetary that is suitable for small telescopes is **IC 4663**, which is a pale grayish color, although some

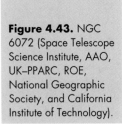

Figure 4.43. NGC 6072 (Space Telescope Science Institute, AAO, UK–PPARC, ROE, National Geographic Society, and California Institute of Technology).

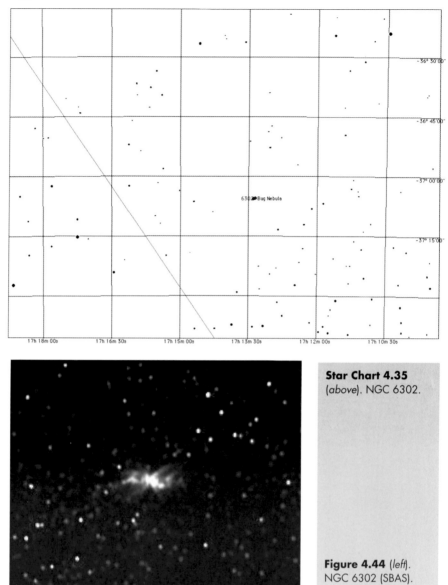

6302 Bug Nebula

-36° 30'00"
-36° 45'00"
-37° 00'00"
-37° 15'00"

17h 18m 00s   17h 16m 30s   17h 15m 00s   17h 13m 30s   17h 12m 00s   17h 10m 30s

Star Chart 4.35
(*above*). NGC 6302.

Figure 4.44 (*left*).
NGC 6302 (SBAS).

observers report it as being bluish. At 20 arcseconds across it should be visible in a 10 cm telescope as a pale circular haze.

There are a few emission nebulae in Scorpius, but most of them require a large-aperture telescope in order for any detail to be seen. We will look at just a few. The first is **IC 4628**, which we mentioned earlier when we discussed the open cluster Trumpler 24. It is faint and so will benefit from the use of an [OIII] filter with a 15 cm telescope. The nebula will appear as a 20 × 10 arcsecond pale smudge at one end of the cluster (see Figure 4.46). Our next object, the emission nebula **NGC 6334,** also known as the **Cat's Paw Nebula,** would have been a wonderful object if its light wasn't obscured by the thick clouds of dust that

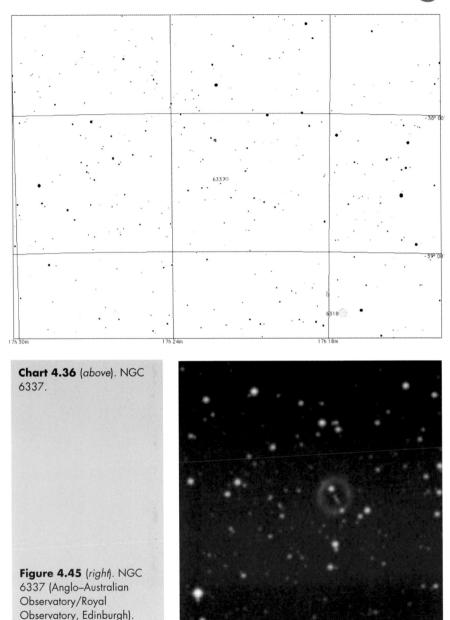

Chart 4.36 (*above*). NGC 6337.

Figure 4.45 (*right*). NGC 6337 (Anglo–Australian Observatory/Royal Observatory, Edinburgh).

are part of the Milky Way (see Star Chart 4.37). In photographs it appears very red due to the HII emission, but to the naked eye only a small amount of green and blue light manages to reach us and so visually is very faint (see Figure 4.47). Its brightest regions can be glimpsed in a 20 cm telescope, and of course an [OIII] filter will help appreciably. Another nebula that suffers from an appreciable amount of extinction is **NGC 6357**. Lying close to NGC 6334, it suffers about the same amount of extinction and only its brightest areas, about 3 × 1 arcminute, can be observed lying northwest of a line of four bright

Figure 4.46. IC 4628 (Matt BenDaniel, http://starmatt.com).

foreground stars (see Star Chart 4.37). If an [OIII] filter is used, it will be seen that the nebula covers a much larger part of the sky (see Figure 4.47). Incidentally, near the nebulosity is the faint open cluster **Pismus 24.**

I have mentioned above how the dust clouds that lie in the Milky Way obscure many of the objects. In some areas, the dust is so dark it can be seen as a dark area against the brilliance of the Milky Way's stars. Such a dark nebula is **Bernes 149**, which is close to the

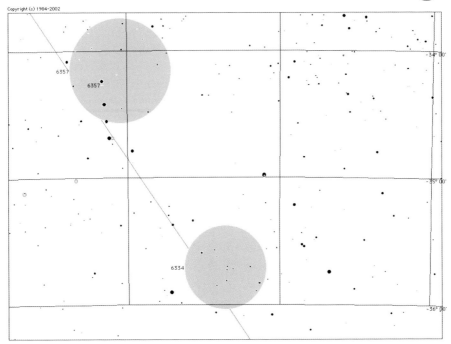

Copyright (c) 1984-2002

Star Chart 4.37. NGC 6334; NGC 6357.

border with Lepus just over 2° southeast of **Eta (η) Scorpii**. It looks like an irregular streak in a star field 1° in length but only 12 arcminutes wide. A dark nebula that really does look like a hole in the Milky Way is **Barnard 50**. It is situated 2.5° east of **Epsilon (ε) Scorpii**. As the Milky Way is particularly rich here, the dark nebula really does stand out.

Probably the most conspicuous dark cloud is **Barnard 283**. This is a very large dust cloud located about 1° northwest of the open cluster M7. It runs from east to west, appearing to begin in a diffuse star field, then slightly thinning as we observe its length, before widening again towards its westernmost point. It can easily be seen in binoculars and small telescopes

Many more dust clouds and star fields and clusters are within this lovely rich region of the Milky Way, but we must leave Scorpius and move on to our penultimate constellation for this section – Ophiuchus.

4.9 Ophiuchus

The Milky Way passes through most, but not all, of this large constellation and this means that although the constellation center transits in June, the parts we are interested in could easily have been placed in Book 1. However, I am sure that when you are scanning Scorpius you will inevitably also look here, so I have decided that this section is where **Ophiuchus** belongs (see Star Chart 4.38).

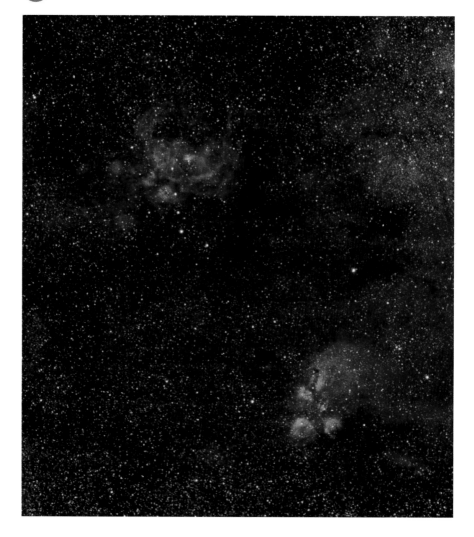

Figure 4.47. NGC 6334; NGC 6357 (Matt BenDaniel, http://starmatt.com).

When we look at this part of the Milky Way, we are peering along the edge regions of our Galaxy. This means that much of the constellation is heavily obscured with dust and gas. However, it is an area rich in deep-sky objects, and has some spectacular globular clusters and dark nebulae, as well as a fine complement of other gems.

Our first object is the star **Rho (ρ) Ophiuchi** in an area that may look pretty ordinary to the naked eye, but it presents an entirely different prospect when photographed. I am sure all of you will have seen the wonderful color images of this region. It is as if someone has taken a rainbow, mixed the colors, then swirled them over the area surrounding Rho Ophiuchi. But I am getting ahead of myself, so back to the star. It is a nice triple star easily seen with a small telescope or large binoculars. The 5th magnitude primary is a blue star set in a close triangle with 7th and 8th magnitude secondaries. In fact, this is a quintuplet

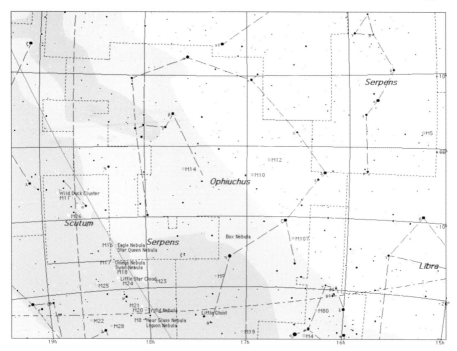

Star Chart 4.38. Ophiuchus.

system, as the primary is itself a very close binary of near equal magnitude stars, separated by about 3 arcseconds. However, the star is immersed in clouds of dust and gas, which although colorful, as I mentioned above, are very faint. The barely luminous haze of reflection nebulae and dark clouds of dust may just be glimpsed in good conditions. The bright[11] areas are known as **IC 4604** and shine by reflecting the starlight from Rho Ophiuchi, whereas the dark clouds are known as **Barnard 42** (see Figure 4.48).

Several other multiple systems can also be found. **Omicron (o) Ophiuchi** is an easy double of magnitudes 5.4 and 6.9, and is located in a field of bright stars. This double makes a nice contrast of orange and yellow. A difficult object, however, is **Lambda (λ) Ophiuchi**. Even though it is a bright object, magnitudes 4.1 and 5.2, it is separated by only 1.4 arcseconds. However, it is a nice pairing of white and pale lemon stars and makes a good test for small telescopes. A larger aperture will show that it is in fact a quadruple system, with the two fainter stars shining at magnitudes 11.1 and 9.5. A lovely double star is **70 Ophiuchi**. It has fine colors of orange and yellow stars and the system is slowly opening; it will be at its widest in 2024 with a separation of 6.8 arcseconds. The colors of these stars, as in a lot of double-star systems, are always open to controversy, as some observers have seen them as yellowish and rose colored. The stars are apparently only 17 light years away. Another beautiful double star is **36 Ophiuchi**, consisting of two lovely orange stars, although some think the secondary has a yellowish tint to it. It is easily resolved in a small telescope. Furthermore, two more stars can be glimpsed: an orange

---

[11] Relatively speaking, of course!

Figure 4.48. IC 4604; Barnard 42 (Chuck Vaughn).

magnitude 8.1 star, some 208 arcseconds to the northwest, and a magnitude 6.6 star 12.2 arcminutes away. All in all it is a lovely system to observe.

There are quite a few variable stars in Ophiuchus, and we will look at just two of these – **Chi (χ) Ophiuchi** and **RS Ophiuchi**. The former is a typical long-period variable star with a magnitude that ranges from 6.0 to 9.3 in 334 days, just short of a year! The latter star is somewhat special in that it is termed a recurrent nova. Normally it shines at magnitude 11.5 and so is very faint; however, it has (1898, 1933, 1958 and 1967) flared up to 4th magnitude. The star is part of a double system, possibly coupled with a white dwarf star, and involves a transfer of material from one to the other, with a resulting increase of brightness.[12] Although we cannot predict what it will happen again with any certainty, it is always worth checking as you could be the first person to see it flare up, and you will become famous and known to every variable-star observer in the world.

A star that should be mentioned, as it is the brightest in the constellation, with a magnitude of 2.08, is **Alpha (α) Ophiuchi**, or **Ras Alhague**. An interesting star for several reasons, it shows the same motions through space as several other stars called the **Ursa**

---

[12] Several of the books listed in the appendices will have fuller and more rigorous descriptions of the mechanisms involved.

Figure 4.49. Messier 107 (Robert Schulz, AAS Gahberg).

Major Group. This moving stream of stars includes the five central stars of the Plough. It is spread over a vast area of the sky, approximately 24°, and is around 20 × 30 light years in extent. It includes as members **Sirius (Alpha (α) Canis Majoris), Alpha (α) Coronae Borealis, Delta (δ) Leonis, Beta (β) Eridani, Delta (δ) Aquarii** and **Beta (β) Serpentis**. Due to the predominance of A1 and A0 stars within the association, its age has been estimated at 300 million years. Ras Alhague also shows interstellar absorption lines in its spectrum. Finally, measurements show an oscillation, or wobble, in its proper motion, which would indicate an unseen companion star.

Our final star is a very famous one, **Barnard's Star (GL 699)**. It is the third-closest star to the Solar System, shining at magnitude 9.54, and is an M-type red dwarf star. It also has the largest proper motion of any star, 10.3 arc seconds per year due north. Thus it would take about 180 years for the star to move the distance equivalent to the Moon's diameter across the sky. The star will appear as a red-tinted point of light some 3.5° east of **Beta (β) Ophiuchi**. It is a worthwhile project for amateurs to observe the star with a medium aperture and high magnification, note its position, and to repeat the observation over several years. You will see that it moves perceptibly against the star background.

Let's now look at those objects for which Ophiuchus is rightly famous – dark nebulae and globular clusters.[13]

Our first globular cluster is **Messier 107 (NGC 6171)**. It is often missed off amateurs' observing schedules owing to its faintness at magnitude 8.1. However, it is nevertheless a pleasant cluster with a mottled disk and brighter core (see Figure 4.49).

Not visible with the naked eye, it nevertheless presents a pleasing aspect when medium to high magnification is used, and its outlying members can be resolved with a 15 cm telescope. A smaller telescope, say 8 cm, will only show a faint patch of haze (see Star Chart 4.39). What makes this inconspicuous globular important, however, is that it is one of the very few that seem to be affected by the presence of interstellar dust. Deep imaging has revealed several obscured areas within the cluster, possibly due to dust grains lying between us. This isn't such a surprise, as the globular is located over the hub of the Galaxy in Scorpius.

---

[13] There are a lot of globular clusters in Ophiuchus and we shall just look at the brighter ones.

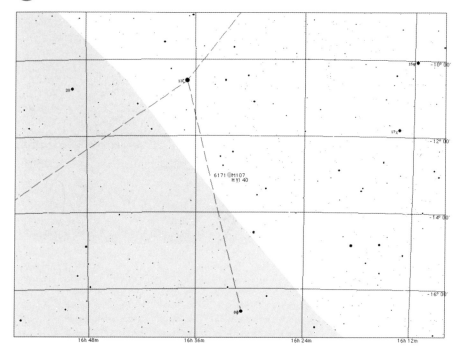

Star Chart 4.39 (*above*). Messier 107.
Star Chart 4.40 (*below*). Messier 10; Messier 12.

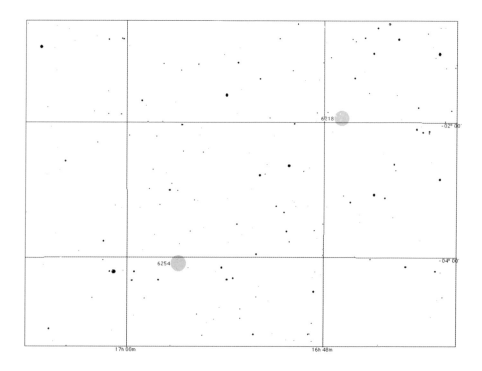

Figure 4.50. Messier 12 (Klaus Eder and Harald Strauss, AAS Gahberg).

Then there is **Messier 12 (NGC 6218)**. This is a small cluster that will be a challenge to naked-eye observers (see Star Chart 4.40). A telescope of 8 cm will show a few stars, whereas in telescopes of aperture 20 cm and more, this cluster will be very impressive, with many stars being resolved against the fainter background of unresolved members (see Figure 4.50). It also contains many faint colored stars which show up well with telescopes of aperture 10 cm and greater.

It is nearly the twin of **Messier 10 (NGC 6254)**, which is within 3° southeast, and so can be seen within the same field of view in a low-power eyepiece (see Figure 4.51). Similar to M12, M10 is, however, slightly brighter and more concentrated. It can be easily seen with the naked eye on dark nights. It lies close to the orange star **30 Ophiuchi** (spectral type K4, magnitude 5), and so if you locate this star, then by using averted vision M10 should be easily seen (see Star Chart 4.40). With apertures of 20 cm and more, the stars are easily resolved right to the cluster's center. Under medium aperture and magnification, several colored components have been reported: a pale blue tinted outer region surrounding a very faint pink area, with a yellow star at the cluster's center. Although M10 and M12 look similar, their structures are in fact quite different. M10 has a moderate condensation of stars, and has an obviously brighter core that can even be seen in binoculars, whereas M12 has a looser structure with a less obvious core.

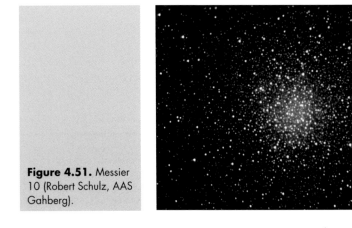

Figure 4.51. Messier 10 (Robert Schulz, AAS Gahberg).

Figure 4.52. Messier 62 (Space Telescope Science Institute, AAO, UK–PPARC, ROE, National Geographic Society, and California Institute of Technology).

One of the less-concentrated types of cluster is **NGC 6235**. This is a hazy bright globular about 1.5 arcminutes across. In small telescopes it will only look like a pale blob, and a 20 cm telescope is needed to see any stars at all, but nevertheless it is a pleasing object to the eye. On the other hand, the brightest globular in Ophiuchus is a delight: **Messier 62** (**NGC 6266**). Shining at magnitude 6.7 and lying about 7° southeast of Antares, this is a very nice cluster, visible in binoculars as a small hazy patch of light set in a wonderful star field (see Figure 4.52). Most globular clusters seem to be above, or below, the plane of the Milky Way, and yet M62 seems to lie within it. This may be the cause of its irregular shape, which bears a cometary appearance, which is apparent even in small telescopes. It has a very interesting structure where concentric rings of stars have been reported by several observers, along with a colored sheen to its center, described as both pale red and yellow (see Star Chart 4.41). Recent work indicates that there are over 80 variable stars within the cluster, which apparently is a lot, most being short-period **RR-Lyrae** stars.

Another fine cluster is **Messier 19** (NGC 6273). Located about 4.5° north of Messier 62, this is a splendid, albeit faint, cluster when viewed through a telescope. It nevertheless can be glimpsed with binoculars, where its egg shape is very apparent (see Figure 4.53). Although a challenge to resolve, it is nevertheless a colorful object, reported as having both faint orange and faint blue stars, while the overall color of the cluster is a creamy white. With binoculars it will resemble a hazy spot, while in small telescopes of 10 cm a distinct granular appearance is evident, along with a few stars (see Star Chart 4.42). Some amateurs also claim that a few faint dark patches mottle the cluster; perhaps this is interstellar dust between us and the cluster. A somewhat difficult object is the globular **NGC 6284**, which can just be seen with a 10 cm telescope, but it will really need an aperture of 20 cm in order for any resolution to be achieved. The cluster is apparently located on the far side of the galactic center (see Star Chart 4.42).

Another small globular is **NGC 6287,** which lies in a part of the Milky Way that has a considerable amount of dark dust and so obscures the cluster. It is a faint cluster with an integrated magnitude of about 8 and is around 1.5 arcminutes across. A 10 cm telescope will show a small spot, whereas in a 30 cm telescope some stars will be just resolved. A similar situation also exists for **NGC 6316** and **NGC 6304 (Herschel 147)**, also immersed in the Milky Way and subject to a considerable amount of interstellar absorption. The former is a small hazy spot about 1 arcminute across that will take a very large telescope to be

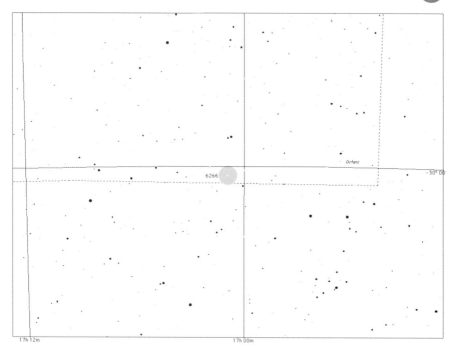

Star Chart 4.41 (*above*). Messier 62.
Star Chart 4.42 (*below*). Messier 19; NGC 6284.

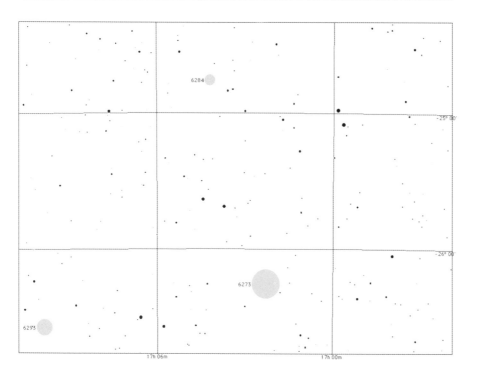

Figure 4.53. Messier 19 (Space Telescope Science Institute, AAO, UK–PPARC, ROE, National Geographic Society, and California Institute of Technology).

resolved. Even a 30 cm aperture does not show any stars. The latter is a nearby cluster but like NGC 6316 it is small but bright, with only a few resolvable stars near its edge. Nevertheless it will be a challenge to locate with binoculars, and will need a 30 cm telescope at least to show individual stars (see Star Chart 4.43).

Star Chart 4.43. NGC 6316; NGC 6304.

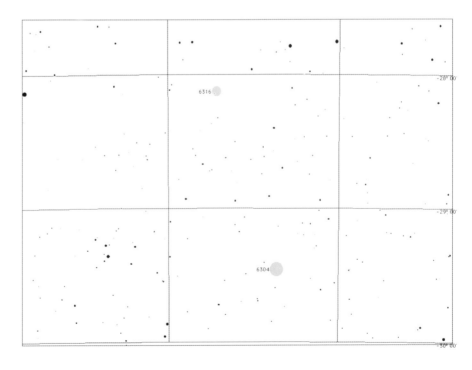

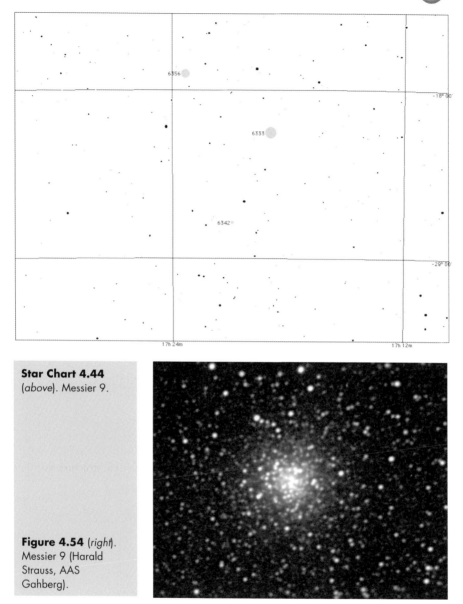

Star Chart 4.44 (*above*). Messier 9.

Figure 4.54 (*right*). Messier 9 (Harald Strauss, AAS Gahberg).

Visible in binoculars is **Messier 9 (NGC 6333)** (see Star Chart 4.44). This is a small globular, with a brighter core. The cluster is one of the nearest to the center of our Galaxy, and is in a region conspicuous for its dark nebulae, including **Barnard 64**; it may be that the entire region is swathed in interstellar dust, which gives rise to the cluster's dim appearance (see Figure 4.54). The dark nebula, Barnard 64, can only be glimpsed when the darkest skies are available so that its contrast to the background Milky Way is at a maximum. It lies about 19,000 light years away.

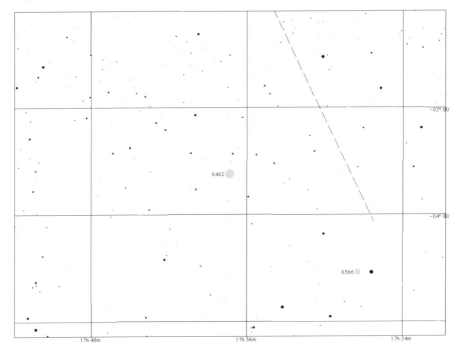

Star Chart 4.45. Messier 14.

A nice cluster is **Messier 14** (**NGC 6402**). Located in an empty part of the sky, it is brighter and larger than is usual for a globular (see Star Chart 4.45). Though visible only in binoculars as a small patch of light, and not resolved even in a small telescope (<15 cm), it is nevertheless worth searching for (see Figure 4.55). It shows a delicate structure with a lot

Figure 4.55. Messier 14 (Klaus Eder and Harald Strauss, AAS Gahberg).

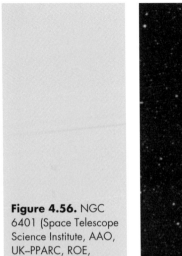

Figure 4.56. NGC 6401 (Space Telescope Science Institute, AAO, UK–PPARC, ROE, National Geographic Society, and California Institute of Technology).

of detail, much of which will be obscured if seen from an urban location. It has a pale yellow tint, and some observers report seeing a definite stellar core, which has a striking orange color. But this feature is seen only with telescopes of aperture 15 cm and greater and using a high magnification.

There are several small and faint globulars in Ophiuchus that are often passed over, but are worthwhile observing, so let's look at them. The first is **NGC 6342**. This is a faint and rather small object at magnitude 9.8 and about 40 arcseconds across. It is unresolvable even in a 30 cm telescope. Another is **NGC 6356**, which is brighter than its predecessor, some 3 arcminutes across, but still remains unresolved in small and medium telescopes. A third cluster is **NGC 6366**, which also suffers from considerable obscuration from dust. It is very faint and has an irregular shape. It can be glimpsed as a dim hazy spot in a 15 cm telescope but should show some resolution in a 25 cm telescope.

Located within a very nice part of the Milky Way is **NGC 6401**. Although it can be seen in a 15 cm telescope, it will remain unresolved, and in larger telescopes will show a definite elliptical shape (see Figure 4.56). Our final globular, **NGC 6426**, is a very distant cluster and as such gives rise to its faintness. In a telescope all you will see is a nice star field and a tiny blob 1 arcsecond across. It will need perseverance to find in a telescope of 15 cm aperture, and larger apertures will show a few stars. It is estimated to lie 17,000 parsecs away.

The Milky Way here is resplendent with many dark clouds. In fact, there may be more observable dark clouds here than anywhere else. There is always one problem associated with these objects, however, as one must have dark skies. So if the Moon is about, or the transparency is poor, then you won't find them. Having said that, let's look at a few. Our first object is the **Pipe Nebula**, which is an enormous object that extends for over 7° in the southern part of Ophiuchus. It is so dark that it can be glimpsed against the starry background even with the naked eye. It can be seen to be in several parts. One area is called **Barnard 78 (LDN 42)** and is also known as the **Pipe Nebula (Bowl)**. Part of the same dark nebula as above, the bowl appears as a jagged formation, covering over 9°. The whole region is studded with dark nebulae, and is thought to be a part of the same complex as that which encompasses Rho (ρ) Ophiuchi and Antares, which are over 700 light years

Figure 4.57. Barnard 72 (Chuck Vaughn).

away from it. Another part is **Barnard 59, 65–7**, also known as the **Pipe Nebula (Stem),** and **Lynds Dark Nebula 1773.** It is best viewed with lower-power binoculars, With the unaided eye, it appears as a straight line, but under magnification its many variations can be glimpsed. It extends for over 5° to the west from the base of the bowl. Our final dark cloud is **Barnard 72**, also known as the **Snake Nebula**. This is a familiar object as it has often been photographed (see Figure 4.57). The background star field is less bright here than for the other dark nebulae, and so it is sometimes a difficult object to locate. The nebula is about 20 arcminutes across and is at its widest around its southeast loop. Large binoculars or a rich-field telescope are the preferred equipment with which to see this object. They will increase the clouds' contrast against the stars.

In addition to the many fine globular clusters we discussed earlier, there are only a few open clusters we can observe. One of these is **IC 4665**. This large cluster is a naked-eye object under perfect seeing conditions, and will appear as a hazy spot measuring over two full moon diameters. With binoculars, nearly 30 blue-white 6th magnitude stars can be seen. Its position in a sparse area of the sky emphasizes the cluster, even though it is not a particularly dense collection of stars (see Star Chart 4.46). Another fine open cluster is **Herschel 72 (NGC 6633)**. Bordering on naked-eye visibility, this bright, large but loose cluster is perfect for binoculars and small telescopes (see Figure 4.58). It is about 30 arcminutes across, shining at 4th magnitude. The stars are a lovely bluish-white set against the faint glow of the unresolved members. At the northern periphery of the cluster is a small but nice triple star system.

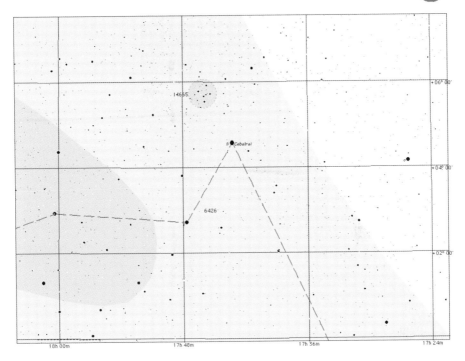

Star Chart 4.46. IC 4665.

An often overlooked cluster[14] is **Collinder 350**. It is another large cluster located near **Beta (β) Ophiuchi**. It only has around 25 stars that really take some effort to form a cluster. To distinguish it from being just a run-of-the-mill collection of field stars, look for four stars of 8th magnitude that make an arc running from north to south pointing to

Figure 4.58. NGC 6633 (Rolf Löhr, AAS Gahberg).

---

[14] I use the word "cluster" advisedly here.

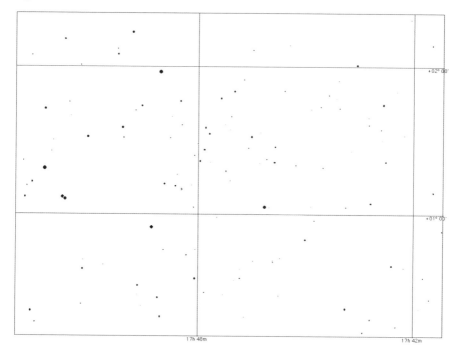

Star Chart 4.47. Collinder 350.

another arc of three stars of 8th magnitude aligned east–west (see Star Chart 4.47). You can be excused for thinking it is not a cluster.

Another nice but faint cluster is **Trumpler 26 (Harvard 15)**. It can be found some 25 arcminutes northeast of **45 Ophiuchi**, and in small telescopes of, say, 10–15 cm aperture, is a 7 arcminute patch of 15 bright stars and about 30 fainter stars.

It is unfortunate that one of the largest clusters in the sky is also one of the most difficult to observe. The cluster in question is **Melotte 186**. The group's center lies more or less directly on the star **67 Ophiuchi**. The boundary of the cluster is some 2° away. The problem comes when you have to decide what star is a member of the cluster and what stars are background objects. Not an easy task. However, if you use low-power binoculars it is possible to see that it does exist, as many stars from 4th to 8th magnitude become transformed into a cluster.

Our final objects are planetary nebulae. There are a respectable number in Ophiuchus, though most are faint and small. There are two exceptions: NGC 6572 and NGC 6369.

The planetary **NGC 6572** is a bright 8th magnitude and distinctly bluish[15] object that shows a definite disk, some 15 arcseconds in diameter. It is easily found in a small telescope, say 10 cm, and in fact can be seen in a finder scope if it is fitted with an [OIII] filter (see Star Chart 4.48). With a high enough aperture, say 30 cm and greater, the central star can be glimpsed. The planetary **NGC 6369**, also known as the **Little Ghost Nebula**, is located in the bowl of the Pipe Nebula, Barnard 78 (see Star Chart 4.49). In a sufficiently large telescope under high magnification, a definite ring or annular shape can be seen,

---

[15] Some observers report that it is distinctly greenish in color.

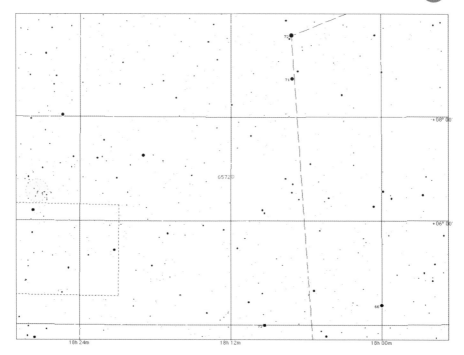

Star Chart 4.48 (*above*). NGC 6572.
Star Chart 4.49 (*below*). NGC 6369.

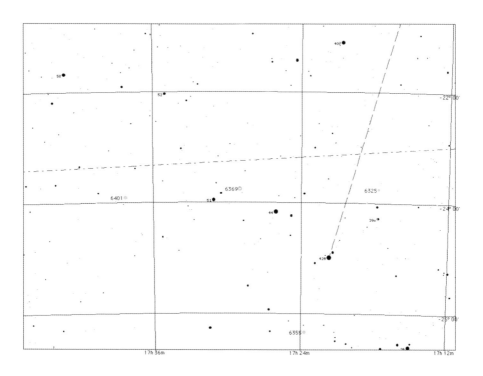

Figure 4.59. NGC 6369 (TLRBSE01/ Adam Block/AURA/ NOAO/NSF).

although it is much fainter than its northern cousin, M57 in Lyra. Measured to be some 30 arcseconds or more in diameter, it can be glimpsed in a small telescope fitted with an [OIII] filter (see Figure 4.59).

4.10 Corona Australis

Our final constellation for this part of the year is not well-known amongst amateurs in the north, yet it is possible to see quite a lot of it from northern latitudes. The constellation **Corona Australis** is located along the southern Milky Way, and offers a wide variety of objects for both binoculars and telescopes (see Star Chart 4.50). It is perfect for scanning with binoculars on early summer evenings, as it contains many splendid star fields and even one obvious dark nebula. It transits on 30 June.

There are several double stars we can look at that range in ease of resolution from very easy to difficult. An easy double is **Kappa (κ) Coronae Austrini**. This is a wide bright pair of stars of magnitudes 6.31 and 5.67, separated by about 21 arcseconds. A close bright double is **Gamma (γ) Coronae Austrini**. The magnitudes of these two stars are 5.03 and 5.10, separated by about 2.6 arcseconds. It is easily split in a telescope of around 10 cm. Our penultimate double star is **h (Herschel) 5014**. This system is composed of nearly equal magnitude stars, namely 5.7 and 5.78. It is a close pair at a separation of about 1.9 arcseconds, so a telescope of 10 cm will be needed in order to resolve the system, and in fact a larger aperture may be needed. Our final double star is **Brs 14**. This is a nice system for small telescopes, consisting of two white stars of magnitudes 6.6 and 6.8 separated by around 12 arcseconds. What makes this pair so interesting is that both stars are spectroscopic binaries and are immersed in a very faint reflection nebula **IC 4812**, which to my knowledge can only be seen photographically or by CCD imaging. Located only 13 arcminutes to the northeast is a much brighter reflection nebula complex, **NGC 6726–27**, and the very strange object NGC 6729, about which we will talk later.

There is a nice globular cluster, **NGC 6541**, which is deep amongst a beautiful starstrewn field and is wonderful to observe. In binoculars it will appear as a just-not-stellar

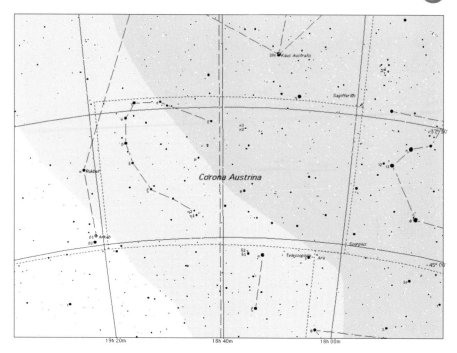

Star Chart 4.50. Corona Australis.

point, but in telescopes it will reveal itself to be a well-condensed round cluster (see Figure 4.60). A small telescope of 10 cm will not resolve its structure but increasing aperture will readily show many sparkling stars up to 6 arcminutes in diameter (see Star Chart 4.51).

The planetary nebula **IC 1297** is a nice object in a telescope of about 15 cm aperture, where it will reveal itself as a pale blue disk some 10 arcseconds in diameter shining at 11th magnitude (see Star Chart 4.52). Much harder to see is the planetary nebula **Fg 3**. Its diameter is only a tiny 2 arcseconds, so it will appear in most instruments as a stellar-like point of light. To resolve it is a challenge indeed!

There are a lot of nebulae, both dark and emission, in Corona Australis, and at several places this material is illuminated by stars, but not just any old stars. Consider the nebula **NGC 6729**. It is located in a blank part of the sky, which indicates immediately that there must be a lot of interstellar dust hereabouts (see Figure 4.61). A few stars are scattered throughout the area and these illuminate an object that looks like a faint comet. It is in fact a reflection nebula, NGC 6729. It is about 1.5 arcminutes in length and has a small star, **R Coronae Austrini,** which incidentally is a protostar, located near its northwest apex (see Star Chart 4.53). What is so peculiar is that the star is a variable, changing its magnitude from 9.7 to 13.5, and the nebula changes in magnitude in time with the star. So we have what can only be termed a variable nebula!

There is another similar object, **NGC 6726** and **6727,** shaped like a figure "8" located to the northwest. It has two stars within the cloud, one of which is another irregular variable star, **TY Coronae Austrini**. The magnitude change for this star is from 8.8 to 12.5, and once again, the nebula varies in magnitude, generally following the changes in the star (see Figure 4.62). Both these nebulous objects are fairly easy to observe with telescopes of, say,

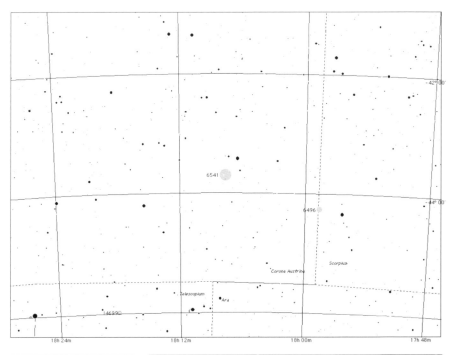

Star Chart 4.51
(*above*). NGC 6541.

Figure 4.60 (*right*).
NGC 6541 (Space
Telescope Science
Institute, AAO,
UK–PPARC, ROE,
National Geographic
Society, and California
Institute of Technology).

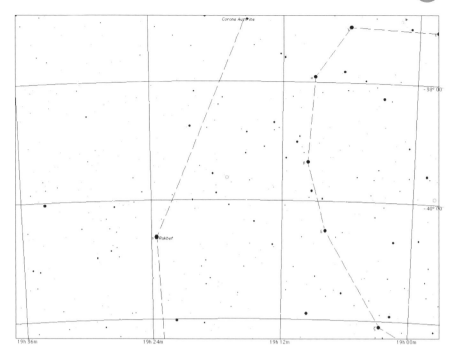

Star Chart 4.52 (*above*). IC 1297.
Star Chart 4.53 (*below*). NGC 6729; NGC 6726; NGC 6727.

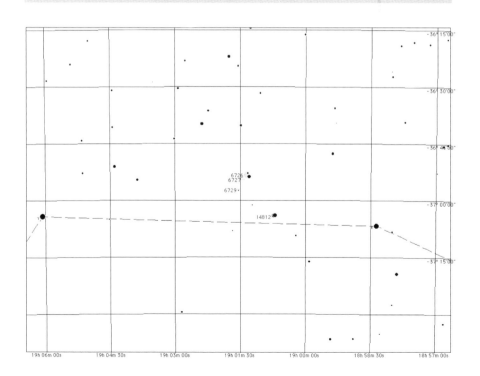

Figure 4.61. NGC 6729 (Space Telescope Science Institute, AAO, UK–PPARC, ROE, National Geographic Society, and California Institute of Technology).

15 cm aperture and greater (see Star Chart 4.53). There is one last point to discuss here relating to these odd objects. The origin of the variability is still open to question, Are the stars inherently variable and thus responsible for the change in nebula brightness, or is there some unseen obscuring material passing in front of the stars which gives the impression of variability? Only time will tell.

Before we leave Corona Australis, let's stay with dark nebulae, and look at **Bernes 157**. This is a very large and irregularly shaped dark cloud that seemingly lies between Gamma (γ) Coronae Austrini and NGC 6726. It can be seen readily, and is in fact best viewed through binoculars when the complete absence of stars will make it impossible to mistake. A somewhat smaller and so less obvious dark nebula lies about 1° to the east of Gamma (γ) Coronae Austrini and has the name **Sandqvist and Lindroos 42**.

The following constellations are also visible during these months at different times throughout the night. Remember that they may be low down and so diminished by the effects of the atmosphere. Also, you may have to observe them either earlier than midnight, or some considerable time after midnight, in order to view them.

Northern Hemisphere

Andromeda, Antila, Aquila Delphinus, Camelopardalis, Cassiopeia, Cepheus, Corona Austrini, Cygnus, Hercules, Lacerta, Lepus, Lyra, Monoceros, Perseus, Puppis, Pyxis, Sagitta, Sagittarius, Scutum, Serpens, Serpens, Vulpecula.

Figure 4.62. NGC 6726 (Matt BenDaniel, http://starmatt.com).

Southern Hemisphere

Canis Major Puppis, Antila, Aquila, Carina, Centaurus, Chamaeleon, Circinus, Columba, Crux, Cygnus, Delphinus, Hercules, Lacerta, Lepus, Lyra, Musca, Octans, Pyxis, Sagitta, Scutum, Serpens Cauda, Telescopium, Puppis, Vela, Sagittarius, Volans, Vulpecula.

Objects in Triangulum Australe

Stars

Designation	Alternate name	Vis. mag	RA	Dec.	Description
I 332		6.8, 8.5	$15^h 20.7^m$	$-67° 29'$	PA 112°; Sep. 0.8"[1]
Slr 11		6.5, 8.8	$15^h 54.9^m$	$-60° 45'$	PA 96°; Sep. 1.2"
Rmk 20		6.2, 6.4	$15^h 47.9^m$	$-65° 27'$	PA 147°; Sep. 1.8"
Iota (ι) Trianguli Australis	Δ 201	5.3, 9.4	$16^h 28.0^m$	$-64° 03'$	PA 12°; Sep. 12.0"
R Trianguli Australis		6.4–6.9	$15^h 19.8^m$	$-66° 30'$	Cepheid variable star
U Trianguli Australis		7.5–8.3	$16^h 07.3^m$	$-62° 55'$	Cepheid variable star
S Trianguli Australis		6.1–6.8	$16^h 01.2^m$	$-63° 47'$	Cepheid variable star

Deep-Sky Objects

Designation	Alternate name	Vis. mag	RA	Dec.	Description
NGC 5979	PK 322–05.1	11.5	$15^h 47.7^m$	$-61° 13'$	Planetary nebula
NGC 5844	He 2–119	12.5	$15^h 10.7^m$	$-64° 40'$	Planetary nebula
NGC 6025	Caldwell 95	5.1	$16^h 03.7^m$	$-60° 30'$	Open cluster

[1] Bear in mind that the position angle and separation may change over a short period of time, and so the values given here may be different from what you observe now.

Objects in Apus

Stars

Designation	Alternate name	Vis. mag	RA	Dec.	Description
Delta (δ) Apodis		4.68, 5.27	$16^h 20.3^m$	−78° 41'	Wide double star
I 236		5.8, 8.0	$14^h 53.2^m$	−73° 11'	PA 121°; Sep. 2.2″
CapO 15		6.9, 8.7	$15^h 06.3^m$	−72° 11'	PA 43°; Sep. 1.6″
Theta (θ) Apodis		6.4–8.6p	$14^h 05.3^m$	−76° 48'	Variable star

Deep-Sky Objects

Designation	Alternate name	Vis. mag	RA	Dec.	Description
IC 4499		10.2	$15^h 00.3^m$	−72° 12'	Globular cluster
NGC 6101		9.2	$16^h 25.8^m$	−76° 48'	Globular cluster
PK 31−13.1	He 2−131	11.6	$15^h 37.2^m$	−71° 55'	Planetary nebula

Objects in Lupus

Stars

Designation	Alternate name	Vis. mag	RA	Dec.	Description
h 4690	Herschel 4690	5.4, 9.2	14^{h}37.3^m	−46° 06'	PA 25°; Sep. 19"
Alpha (α) Lupi	HdO 238	2.3$_v$, 13.4	14^{h}41.9^m	−47° 23'	PA 232°; Sep. 28"
Pi (π) Lupi.	H 4728	4.6, 4.7	15^{h}05.1^m	−47° 03'	PA 69°; Sep. 1.6"
Kappa (κ) Lupi	Δ 177	3.9, 5.8	15^{h}11.9^m	−48° 44'	PA 143°; Sep. 26.7"
Mu (μ) Lupi	h 4753	5.0, 5.1	15^{h}18.5^m	−47° 53'	PA 130°; Sep. 1.1"
Gamma (γ) Lupi	h 4786	3.5, 3.6	15^{h}35.1^m	−41° 10'	PA 282°; Sep. 0.74"
h 4788	Herschel 4788	4.7, 6.6	15^{h}35.9^m	−44° 58'	PA 8°; Sep. 2.2"
Xi (ξ) Lupi		5.1, 5.6	15^{h}56.9^m	−33° 58'	PA 49°; Sep. 10.4"
Tau (τ) Lupi		4.56, 4.93	14^{h}26.1^m	−45° 13'	Wide double star

Deep-Sky Objects

Designation	Alternate name	Vis. mag	RA	Dec.	Description
IC 4406	PK 319+15.1	10.2	14^{h}22.4^m	−44° 09'	Planetary nebula
NGC 5873	PK 331+16.1	11.0	15^{h}12.8^m	−38° 08'	Planetary nebula
NGC 5882	PK 327+10.1	9.4	15^{h}16.8^m	−45° 39'	Planetary nebula
NGC 6026	PK 341+13.1	12.9	16^{h}01.4^m	−34° 32'	Planetary nebula
NGC 5824		9.0	15^{h}04.0^m	−33° 04'	Globular cluster
NGC 5986		7.6	15^{h}46.1^m	−37° 47'	Globular cluster
NGC 5927		8.0	15^{h}28.0^m	−50° 40'	Globular cluster
NGC 5822		5.5p	15^{h}05.3^m	−54° 21'	Open cluster
NGC 5749		8.8p	14^{h}48.9^m	−54° 31'	Open cluster
Barnard 228			15^{h}45.5^m	−34° 24'	Dark nebula

Objects in Norma

Stars

Designation	Alternate name	Vis. mag	RA	Dec.	Description
Iota[1] (ι) Normae	See 258	5.3, 5.5, 8.1	$16^h 03.5^m$	–57° 47'	PA 244°; Sep. 0.2–0.8, 11"
Epsilon (ε) Normae	h 4853	4.8, 7.5	$16^h 27.2^m$	–47° 33'	PA 334°; Sep. 23"
h 4813	Herschel 4813	5.9, 8.5	$15^h 55.5^m$	–60° 11'	PA 100°; Sep. 3.9"

Deep-Sky Objects

Designation	Alternate name	Vis. mag	RA	Dec.	Description
NGC 5946		9.7	$15^h 35.5^m$	–50° 40'	Globular cluster
NGC 5999		9.0p	$15^h 52.2^m$	–56° 28'	Open cluster
NGC 6067		5.6	$16^h 13.2^m$	–54° 13'	Open cluster
NGC 6087		5.3	$16^h 18.9^m$	–57° 54'	Open cluster
NGC 6134		7.2	$16^h 27.7^m$	–49° 09'	Open cluster
NGC 6152		8.1p	$16^h 32.7^m$	–52° 37'	Open cluster
Harvard 10			$16^h 19.9^m$	–54° 59'	Open cluster
Collinder 299		6.9p	$16^h 18.4^m$	–57° 07'	Open cluster
Ruprecht 113			$15^h 57.2^m$	–59° 28'	Open cluster
Menzel 2	PK 329–02.2	12.0	$16^h 14.5^m$	–54° 57'	Planetary nebula
Shapley 1	PK 329+02.1	12.6	$15^h 51.7^m$	–51° 32'	Planetary nebula
NGC 6164-65		–	$16^h 34.0^m$	–48° 06'	Bipolar emission nebula

Objects in Ara

Stars

Designation	Alternate name	Vis. mag	RA	Dec.	Description
R Arae	h 4866	6.8, 7.9	$16^h 39.7^m$	$-57° 00'$	PA 122°; Sep. 3.6"
CorO 201		7.1, 7.3	$16^h 50.6^m$	$-50° 003'$	PA 42°; Sep. 3.0"
Brs 13	R 297	5.5, 8.5	$17^h 19.1^m$	$-46° 38'$	PA 248°; Sep. 8.3"
Gamma (γ) Arae	h 4942	3.3, 10.5	$17^h 25.4^m$	$-56° 23'$	PA 328°; Sep. 17.9"
h 4949	Herschel 4949	5.7, 6.5	$17^h 26.9^m$	$-45° 051'$	PA 253°; Sep. 2.1"
I 40		6.1, 10.3	$17^h 31.8^m$	$-46° 002'$	PA 210°; Sep. 18.0"
h 4978	V 539 Ara	5.7, 6.4	$17^h 50.5^m$	$-53° 37'$	PA 269°; Sep. 12.3"

Deep-Sky Objects

Designation	Alternate name	Vis. mag	RA	Dec.	Description
NGC 6326	PK 338–08.1	12.2	$17^h 20.8^m$	$-51° 45'$	Planetary nebula
Shapley 3	PK 342–14.1	11.9p	$18^h 07.3^m$	$-51° 02'$	Planetary nebula
NGC 6352	Caldwell 81	7.8	$17^h 25.5^m$	$-48° 25'$	Globular cluster
NGC 6362.		7.6	$17^h 31.9^m$	$-67° 03'$	Globular cluster
NGC 6397		5.8	$17^h 40.7^m$	$-53° 40'$	Globular cluster
NGC 6200		7.4	$16^h 44.2^m$	$-47° 29'$	Open cluster
NGC 6204		8.2	$16^h 46.5^m$	$-47° 01'$	Open cluster
NGC 6208		7.2	$16^h 49.5^m$	$-53° 49'$	Open cluster
IC 4651		6.9	$17^h 24.7^m$	$-49° 57'$	Open cluster
NGC 6193		5.2	$16^h 41.3^m$	$-48° 46'$	Open cluster
NGC 6188		–	$16^h 40.5^m$	$-48° 47'$	Dark nebula
NGC 6221		10.1	$16^h 52.8^m$	$-59° 13'$	Spiral galaxy

Objects in Pavo

Stars

Designation	Alternate name	Vis. mag	RA	Dec.	Description
Lambda (λ) Pavonis		3.4–4.4	$18^h 52.2^m$	–62° 11'	Variable star
Kappa (κ) Pavonis		3.9–4.8	$18^h 56.9^m$	–67° 14'	Variable star
Xi (ξ) Pavonis	Gale 2	4.4, 9.2	$18^h 23.2^m$	–61° 30'	PA 155°; Sep. 3.4"
V Pavonis		6.65_v	$17^h 43.3^m$	–57° 43'	Variable star

Deep-Sky Objects

Designation	Alternate name	Vis. mag	RA	Dec.	Description
IC 4662		11.3	$17^h 47.1^m$	–64° 38'	Dwarf galaxy
NGC 6630		13.9	$18^h 32.6^m$	–63° 17'	Galaxy
NGC 6684		10.5	$18^h 49.0^m$	–65° 11'	Galaxy

Objects in Libra

Deep-Sky Objects

Designation	Alternate name	Vis. mag	RA	Dec.	Description
NGC 5897		8.4	$15^h 17.4^m$	–21° 01'	Globular cluster
Me 2–1	PK 342+27.1	11.6	$15^h 22.3^m$	–23° 38'	Planetary nebula
NGC 5898	Herschel 138	11.4	$15^h 18.2^m$	–24° 06'	Galaxy
NGC 5903	Herschel 139	11.1	$15^h 18.6^m$	–24° 04'	Galaxy

Objects in Scorpius

Stars

Designation	Alternate name	Vis. mag	RA	Dec.	Description
Alpha (α) Scorpii	Antares	1.2, 5.4	16ʰ 29.4ᵐ	–26° 26'	PA 275°; Sep. 2.9"
Beta (β) Scorpii.		2.6, 4.9	16ʰ 05.4ᵐ	–19° 48'	PA 21°; Sep. 13.6"
Nu (ν) Scorpii		4.3, 6.8/6.4, 7.8	16ʰ 12.0ᵐ	–19° 28'	PA 03°; Sep. 0.9"ᴬᴮ PA 51°; Sep. 2.3"ᶜᴰ
2 Sco	ADS 9823	4.7, 7.1	15ʰ 53.6ᵐ	–25° 20'	PA 273°; Sep. 2.3"
HN 39	h 4850	5.8, 6.4	16ʰ 24.7ᵐ	–29° 42'	PA 356°; Sep. 4.8"
Sigma (σ) Scorpii		2.9v, 8.4	16ʰ 21.2ᵐ	–25° 36'	PA 273°; Sep. 20.0"
I 576		5.18, 12.8	16ʰ 55.0ᵐ	–41° 09'	PA 265°; Sep. 5.4"
Howe 86		6.8, 8.9	17ʰ 13.9ᵐ	–38° 18'	PA 144°; Sep. 2.8"
Zeta¹⁺² (ζ) Scorpii		4.8, 6.2	16ʰ 54.3ᵐ	–42° 20'	Wide double star
Mu¹⁺² (μ) Scorpii		3.0, 3.56	16ʰ 51.9ᵐ	–38° 02'	Wide double star
Xi (ξ) Scorpii		4.70, 3.62	16ʰ 54ᵐ	–42° 41'	Wide double star
RR Scorpii		6.97	16ʰ 56.6ᵐ	–30° 44'	Variable star
BM Scorpii		7.00	16ʰ 56.6ᵐ	–30° 34'	Variable star

Deep-Sky Objects

Designation	Alternate name	Vis. mag	RA	Dec.	Description
NGC 6231	Caldwell 76	2.6	16ʰ 54.0ᵐ	–41° 48'	Open cluster
NGC 6124	Caldwell 75	5.8	16ʰ 25.6ᵐ	–40° 40'	Open cluster
NGC 6242		6.4	16ʰ 54.6ᵐ	–39° 30'	Open cluster
NGC 6281		5.4	17ʰ 04.8ᵐ	–37° 54'	Open cluster
NGC 6322		6.0	17ʰ 18.5ᵐ	–42° 57'	Open cluster
NGC 6405	Messier 6 /Butterfly Cluster	4.2	17ʰ 40.1ᵐ	–32° 13'	Open cluster
NGC 6451		8.2p	17ʰ 50.7ᵐ	–30° 13'	Open cluster
Collinder 316		3.4p	16ʰ 55.5ᵐ	–40° 50'	Open cluster
Trumpler 24		8.6p	16ʰ 57.0ᵐ	–40° 40'	Open cluster
NGC 6475	Messier 7	3.3	17ʰ 53.9ᵐ	–34° 49'	Open cluster

NGC 6453		5.4	17ʰ 50.9ᵐ	−34° 36'	Globular cluster
NGC 6093	Messier 80	7.3	16ʰ 17.0ᵐ	−22° 59'	Globular cluster
NGC 6121	Messier 4	5.8	16ʰ 23.6ᵐ	−26° 32'	Globular cluster
NGC 6144	Herschel 10	9.0	16ʰ 27.3ᵐ	−26° 02'	Globular cluster
NGC 6139		8.9	16ʰ 27.7ᵐ	−38° 51'	Globular cluster
NGC 6256		11.3	16ʰ 59.5ᵐ	−37° 07'	Globular cluster
NGC 6380	Ton 1	11.5	17ʰ 34.4ᵐ	−39° 04'	Globular cluster
Ton 2	Pis 26	12.2	17ʰ 36.7ᵐ	−38° 33'	Globular cluster
NGC 6388		6.7	17ʰ 36.3ᵐ	−44° 44'	Globular cluster
NGC 6441		7.2	17ʰ 50.2ᵐ	−37° 03'	Globular cluster
NGC 6453		9.8	17ʰ 50.9ᵐ	−34° 36'	Globular cluster
NGC 6496		9.0	17ʰ 58.0ᵐ	−39° 08'	Globular cluster
NGC 6072	PK 342+10.1	11.7	16ʰ 13.0ᵐ	−36° 14'	Planetary nebula
NGC 6153	PK 341+5.1	10.9	16ʰ 31.5ᵐ	−40° 15'	Planetary nebula
NGC 6302	Bug Nebula/PK 341+5.1	9.6	17ʰ 13.7ᵐ	−37° 06'	Planetary nebula
NGC 6337	PK 349−1.1	12.3	17ʰ 22.3ᵐ	−38° 29'	Planetary nebula
IC 4663	PK 346−08.1	12.5	17ʰ 45.5ᵐ	−44° 54'	Planetary nebula
IC 4628		–	16ʰ 57.0ᵐ	−40° 20'	Emission nebula
NGC 6334		–	17ʰ 20.4ᵐ	−35° 51'	Emission nebula
NGC 6357		–	17ʰ 24.7ᵐ	−34° 12'	Emission nebula
Barnard 50		–	17ʰ 02.9ᵐ	−34° 24'	Dark nebula
Barnard 283		–	17ʰ 51.3ᵐ	−33° 53'	Dark nebula

Objects in Ophiuchus

Stars

Designation	Alternate name	Vis. mag	RA	Dec.	Description
Rho (ρ) Ophiuchi		5.3, 6.0	$16^h 25.6^m$	−23° 27'	PA 344°; Sep. 3.1"
Omicron (o) Ophiuchi		5.4, 6.9	$17^h 18.0^m$	−24° 17'	PA 355°; Sep. 10.3"
Lambda (λ) Ophiuchi.		4.2, 5.2	$16^h 30.9^m$	+01° 59'	PA 30°; Sep. 1.5"
70 Ophiuchi.		4.2, 6.0	$18^h 05.5^m$	+02° 30'	PA 148°; Sep. 3.8"
36 Ophiuchi		5.1, 5.1	$17^h 15.3^m$	−26° 36'	PA 146°; Sep. 4.9"
Chi (χ) Ophiuchi		4.2–5.0	$16^h 27.0^m$	−18° 27'	Variable star
RS Ophiuchi.		4.0–12.0	$17^h 50.2^m$	−06° 43'	Recurrent nova
Alpha (α) Ophiuchi	Ras Alhague	2.08	$17^h 34.9^m$	+12° 33'	Quite a lot actually!
Barnard's Star	GL 699	9.54	$17^h 57.8^m$	+04° 38'	Largest proper motion

Deep-Sky Objects

Designation	Alternate name	Vis. mag	RA	Dec.	Description
NGC 6171	Messier 107	8.1	$16^h 32.5^m$	−13° 03'	Globular cluster
NGC 6218	Messier 12	6.8	$16^h 47.2^m$	−01° 57'	Globular cluster
NGC 6254	Messier 10	6.6	$16^h 57.1^m$	−04° 06'	Globular cluster
NGC 6235	Herschel 584	10.0	$16^h 53.4^m$	−22° 11'	Globular cluster
NGC 6266	Messier 62	5.7	$17^h 01.2^m$	−30° 07'	Globular cluster
NGC 6273	Messier 19	6.7	$17^h 02.6^m$	−26° 16'	Globular cluster
NGC 6284	Herschel 11	8.9	$17^h 04.5^m$	−24° 46'	Globular cluster
NGC 6287	Herschel 195	8.2	$17^h 05.2^m$	−22° 42'	Globular cluster
NGC 6316	Herschel 45	8.8	$17^h 16.6^m$	−28° 08'	Globular cluster
NGC 6304	Herschel 147	8.4	$17^h 14.5^m$	−29° 28'	Globular cluster
NGC 6333	Messier 9	7.9	$17^h 19.2^m$	−18° 31'	Globular cluster
NGC 6402	Messier 14	7.6	$17^h 37.6^m$	−03° 15'	Globular cluster
NGC 6342	Herschel 49	9.8	$17^h 21.2^m$	−19° 35'	Globular cluster
NGC 6356	Herschel 48	8.2	$17^h 23.6^m$	−17° 49'	Globular cluster

NGC 63696		8.9	17^h 24.0^m	-05° 05'	Globular cluster
NGC 6401	Herschel 44	9.5	17^h 38.6^m	-23° 55'	Globular cluster
NGC 6426	Herschel 587	11.1	17^h 44.9^m	-03° 00'	Globular cluster
Barnard 78	LDN 42	–	17^h 33.0^m	-26° 30'	Dark nebula
Barnard 59, 65–7		–	17^h 21.0^m	-27° 23'	Dark nebula
Barnard 72	Snake Nebula	–	17^h 23.5^m	-23° 28'	Dark nebula
IC 4665		4.2	17^h 46.3^m	+05° 43'	Open cluster
NGC 6633	Herschel 72	4.6	18^h 27.7^m	+06° 34'	Open cluster
Collinder 350		6.1p	17^h 48.1^m	+01° 18'	Open cluster
Trumpler 26	Harvard 15	9.5p	17^h 28.5^m	-29° 29'	Open cluster
Melotte 186		3.0p	18^h 01^m	+03°	Open cluster
NGC 6572	PK 34+11.1	8.1	18^h 12.1^m	+06° 51'	Planetary nebula
NGC 6369	Little Ghost Nebula	11.4	17^h 29.3^m	-23° 46'	Planetary nebula

Objects in Corona Australis

Stars

Designation	Alternate name	Vis. mag	RA	Dec.	Description
Kappa (κ) Coronae Austrini		4.8, 5.1	$18^h 33.4^m$	$-38° 44'$	PA 55°; Sep. 1.3"
Gamma (γ) Coronae Austrini		5.9, 6.6	$18^h 33.4^m$	$-38° 44'$	PA 359°; Sep. 21.4"
h 5014	Herschel 5014	5.7, 5.7	$18^h 06.8^m$	$-43° 25'$	PA 345°; Sep. 0.9"
Brs 14		6.6, 6.8	$19^h 01.1^m$	$-37° 04'$	PA 281°; Sep. 12.7"

Deep-Sky Objects

Designation	Alternate name	Vis. mag	RA	Dec.	Description
NGC 6541		6.1	$18^h 08.0^m$	$-43° 42'$	Globular cluster
IC 1297	PK 358–21.1	10.7	$19^h 17.4^m$	$-39° 37'$	Planetary nebula
NGC 6729		–	$19^h 01.9^m$	$-36° 57'$	Emission nebula
NGC 6726–6727		–	$19^h 01.7^m$	$-36° 53'$	Reflection nebula
Bernes 157		–	$19^h 02.9^m$	$-37° 08'$	Dark nebula

Appendix 1
Astronomical Coordinate Systems

A basic requirement for studying the heavens is determining where in the sky things are. To specify sky positions, astronomers have developed several coordinate systems. Each uses a coordinate grid projected on to the celestial sphere, in analogy to the geographic coordinate system used on the surface of the Earth. The coordinate systems differ only in their choice of the fundamental plane, which divides the sky into two equal hemispheres along a great circle (the fundamental plane of the geographic system is the Earth's equator). Each coordinate system is named for its choice of fundamental plane.

The Equatorial Coordinate System

The equatorial coordinate system is probably the most widely used celestial coordinate system. It is also the most closely related to the geographic coordinate system, because they use the same fundamental plane and the same poles. The projection of the Earth's equator onto the celestial sphere is called the celestial equator. Similarly, projecting the geographic poles on to the celestial sphere defines the north and south celestial poles.

However, there is an important difference between the equatorial and geographic coordinate systems: the geographic system is fixed to the Earth; it rotates as the Earth does. The equatorial system is fixed to the stars, so it appears to rotate across the sky with the stars, but of course it's really the Earth rotating under the fixed sky.

The latitudinal (latitude-like) angle of the equatorial system is called **declination** (Dec for short). It measures the angle of an object above or below the celestial equator. The longitudinal angle is called the **right ascension** (RA for short). It measures the angle of an object east of the vernal equinox. Unlike longitude, right ascension is usually measured in hours instead of degrees, because the apparent rotation of the equatorial coordinate system is closely related to sidereal time and hour angle. Since a full rotation of the sky takes 24 hours to complete, there are (360 degrees/24 hours) = 15 degrees in one hour of right ascension.

This coordinate system is illustrated in Figure A.1

The Galactic Coordinate System

The galactic coordinate system has latitude and longitude lines, similar to what you are familiar with on Earth. In the galactic coordinate system the Milky Way uses as its fundamental plane the zero

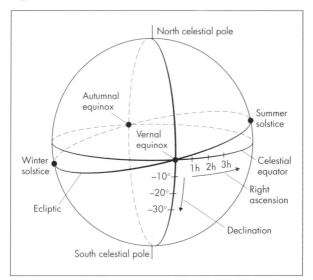

Figure A.1. The equatorial coordinate system.

degree latitude line in the plane of our Galaxy, and the zero degree longitude line is in the direction of the center of our galaxy. The latitudinal angle is called the galactic latitude, and the longitudinal angle is called the galactic longitude. This coordinate system is useful for studying the Galaxy itself.

The reference plane of the galactic coordinate system is the disk of our Galaxy (i.e. the Milky Way) and the intersection of this plane with the celestial sphere is known as the galactic equator, which is inclined by about 63° to the celestial equator. **Galactic latitude,** b (see Figure A.2) is analogous to declination, but measures distance north or south of the galactic equator, attaining +90° at the north galactic pole (NGP) and –90° at the south galactic pole (SGP).

Galactic longitude, l, is analogous to right ascension and is measured along the galactic equator in the same direction as right ascension. The zero point of galactic longitude is in the direction of the galactic center (GC), in the constellation of Sagittarius; it is defined precisely by taking the galactic longitude of the north celestial pole to be exactly 123°. This somewhat confusing system is best shown by the diagram in Figure A.2.

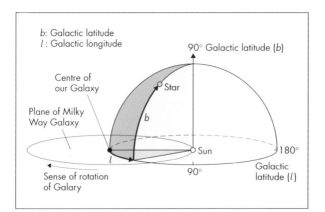

Figure A.2. The Galactic coordinate system.

Appendix 2

Magnitudes

The first thing that strikes even a casual observer is that the stars are of differing brightness. Some are faint, some are bright, and a few are very bright; this brightness is called the magnitude of a star.

The origins of this brightness system are historical, when all the stars seen with the naked eye were classified into one of six magnitudes, with the brightest being called a "star of the first magnitude", the faintest a "star of the sixth magnitude". Since then the magnitude scale has been extended to include negative numbers for the brightest stars, and decimal numbers used between magnitudes, along with a more precise measurement of the visual brightness of the stars. Sirius has a magnitude of –1.44, while Regulus has a magnitude of +1.36. Magnitude is usually abbreviated to m. Note that the brighter the star, the smaller the numerical value of its magnitude.

A difference between two objects of 1 magnitude means that the object is about 2.512 times brighter (or fainter) than the other. Thus a first-magnitude object (magnitude m = 1) is 2.512 times brighter than a second-magnitude object (m = 2). This definition means that a first-magnitude star is brighter than a sixth-magnitude star by the factor 2.512 raised to the power of 5. That is a hundredfold difference in brightness. The naked-eye limit of what you can see is about magnitude 6, in urban or suburban skies. Good observers report seeing stars as faint as magnitude 8 under exceptional conditions and locations. The magnitude brightness scale doesn't tell us whether a star is bright because it is close to us, or whether a star is faint because it's small or because it's distant. All that this classification tells us is the apparent magnitude of the object – that is, the brightness of an object as observed visually, with the naked eye or with a telescope. A more precise definition is the absolute magnitude, M, of an object. This is defined as the brightness an object would have at a distance of 10 parsecs from us. It's an arbitrary distance, deriving from the technique of distance determination known as parallax; nevertheless, it does quantify the brightness of objects in a more rigorous way. For example, Rigel has an absolute magnitude of –6.7, and one of the faintest stars known, Van Biesbroeck's Star, has a value of +18.6.

Of course, the above all assumes that we are looking at objects in the visible part of the spectrum. It shouldn't come as any surprise to know that there are several further definitions of magnitude that rely on the brightness of an object when observed at a different wavelength, or waveband, the most common being the U, B and V wavebands, corresponding to the wavelengths 350, 410 and 550 nanometers respectively. There is also a magnitude system based on photographic plates: the photographic magnitude, m_{pg}, and the photovisual magnitude, m_{pv}. Finally, there is the bolometric magnitude, m_{BOL}, which is the measure of all the radiation emitted from the object.[1]

---

[1] It is interesting to reflect that *no* magnitudes are in fact a true representation of the brightness of an object, because every object will be dimmed by the presence of interstellar dust. All magnitude determinations therefore have to be corrected for the presence of dust lying between us and the object. It is dust that stops us from observing the center of our Galaxy.

Objects such as nebulae and galaxies are **extended objects**, which means that they cover an appreciable part of the sky: in some cases a few degrees, in others only a few arc minutes. The light from, say, a galaxy is therefore "spread out" and thus the quoted magnitude will be the magnitude of the galaxy were it the "size" of a star; this magnitude is often termed the **combined** or **integrated magnitude**. This can cause confusion, as a nebula with, say, a magnitude of 8, will appear fainter than an 8th magnitude star, and in some cases, where possible, the surface brightness of an object will be given. This will give a better idea of what the overall magnitude of the object will be.

Finally, many popular astronomy books will tell you that the faintest, or limiting magnitude, for the naked eye is around the 6th magnitude. This may well be true for those of us who live in an urban location. But the truth of the matter is that from exceptionally dark sites with a complete absence of light pollution, magnitudes as faint as 8 can be seen. This will come as a surprise to many amateurs. Furthermore, when eyes are fully dark-adapted, the technique of averted vision will allow you to see with the naked eye up to three magnitudes fainter! But before you rush outside to test these claims, remember that to see really faint objects, either with the naked eye or telescopically, several other factors such as the transparency and seeing conditions, and the psychological condition of the observer (!) will need to be taken into consideration, with light pollution as the biggest evil.

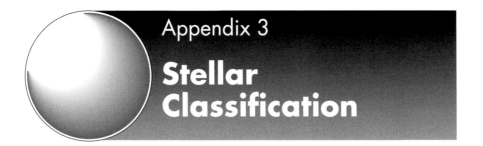

Appendix 3
Stellar Classification

For historical reasons a star's classification is designated by a capital letter thus, in order of decreasing temperature:

O B A F G K M L R N S

The sequence goes from hot blue stars types O and A to cool red stars K and M. and L. In addition, there are rare and hot stars called Wolf–Rayet stars, class WC and WN, exploding stars Q, and peculiar stars, P. The star types R, N and S actually overlap class M, and so R and N have been reclassified as C-type stars, the C standing for carbon stars. A new class has recently been introduced, the L class.[2] Furthermore, the spectral types themselves are divided into ten spectral classes beginning with 0, 1, 2, 3 and so on up to 9. A class A1 star is thus hotter than a class A8 star, which in turn is hotter than a class F0 star. Further prefixes and suffixes can be used to illustrate additional features:

a star with:

emission lines	e (also called f in some O-type stars)
metallic lines	m
a peculiar spectrum	p
a variable spectrum	v
a blue or shift in the line (for example, P-Cygni stars)	q

And so forth. For historical reasons, the spectra of the hotter star types O, A and B are sometimes referred to as **early-type** stars, while the cooler ones, K, M, L, C and S, are later-type. Also, F and G stars are **intermediate-type** stars.

---

[2] These are stars with very low temperatures: 1900–1500 K. Many astronomers believe these are brown dwarves.

Appendix 4

Light Filters

One of the most useful accessories an amateur can possess is one of the ubiquitous optical filters. Having been accessible previously only to the professional astronomer, they came on to the market relatively recently, and have made a very big impact. They are useful, but don't think they're the whole answer! They can be a mixed blessing. From reading some of the advertisements in astronomy magazines you would be correct in thinking that they will make hitherto faint and indistinct objects burst into vivid observability – sometimes!

What the manufacturers do not mention is that regardless of the filter used, you will still need dark and transparent skies for the use of the filter to be worthwhile. Don't make the mistake of thinking that using a filter from an urban location will always make objects become clearer. The first and most immediately apparent item on the downside is that in all cases the use of a filter reduces the total amount of light that reaches the eye, often quite substantially. However, what the filter does do is select specific wavelengths of light emitted by an object, which may be swamped by other wavelengths. It does this by suppressing the unwanted wavelengths. This is particularly effective when observing extended objects such as emission nebulae and planetary nebulae.

In the former case, use a filter that transmits light around the wavelength of 653.2 nm, which is the spectral line of hydrogen alpha (Hα), and is the wavelength of light responsible for the spectacular red color seen in photographs of emission nebulae. Some filters may transmit light through perhaps two wavebands: 486 nm for hydrogen beta[3] (Hβ) and 500.7 nm for oxygen-3 [OIII], two spectral lines which are very characteristic in planetary nebula. Use of such filters will enhance the faint and delicate structure within nebulae, and, from a dark site, they really do bring out previously invisible detail. Don't forget (as the advertisers sometimes seem to) that "nebula" filters do not (usually) transmit the light from stars, and so when in use the background will be dark with only nebulosity visible, and this makes them somewhat redundant for observing stars, star clusters and galaxies alone, unless the aforementioned objects are associated with nebulosity, as can often be the case.

One kind of filter that does help in heavily light-polluted areas is the LPR (light-pollution reduction filter), which effectively blocks out the light emitted from sodium and mercury street lamps, at wavelengths 366, 404.6, 435.8, 546.1, 589.0 and 589.5 nm. Clearly the filter will only be effective if the light from the object you want to see is significantly different from the light-polluting source: fortunately, this is usually the case. Light-pollution reduction filters can be very effective visually and photographically, but remember that there is always some overall reduction in brightness of the object you are observing.

---

[3]This filter can be used to view dark nebulae that are overwhelmed by the proximity of emission nebulae. A case in point is the Horsehead Nebula, which is incredibly faint, and swamped by light from the surrounded emission nebulosity.

Whatever filters you decide on, it is worthwhile trying to use them before you make a purchase (they are expensive!), by borrowing them either from a fellow amateur or from a local astronomical society. This will show you whether the filter really makes any difference to your observing.

There is no doubt that modern filters can be an excellent purchase, but it may be that your location or other factors will prevent the filter from realizing its full potential or value for money. Most commercially available filters are made for use at a telescope and not for binoculars, so unless you are mechanically minded and can make your own filter mounts (and are happy to pay – two LPR filters could easily cost more than the binoculars!), it's likely that only those observers with telescopes can benefit.

Appendix 5

Star Clusters

Open or **galactic clusters**, as they are sometimes called, are collections of young stars, containing anything from maybe a dozen members to hundreds. A few of them, for example, Messier 11 in Scutum, contain an impressive number of stars, equaling that of globular clusters (see below), while others seem little more than a faint grouping set against the background star field. Such is the variety of open clusters that they come in all shapes and sizes. Several are over a degree in size and their full impact can only be appreciated by using binoculars, as a telescope has too narrow a field of view. An example of such a large cluster is Messier 44 in Cancer. Then there are tiny clusters, seemingly nothing more than compact multiple stars, as is the case with IC 4996 in Cygnus. In some cases all the members of the cluster are equally bright, such as Caldwell 71 in Puppis, but there are others that consist of only a few bright members accompanied by several fainter companions, as in the case of Messier 29 in Cygnus. The stars which make up an open cluster are called **Population I** stars, which are metal-rich[4] and usually to be found in or near the spiral arms of the Galaxy.

The reason for the varied and disparate appearances of open clusters is the circumstances of their births. It is the interstellar material out of which stars form that determines both the number and type of stars that are born within it. Factors such as the size, density, turbulence, temperature, and magnetic field all play a role as the deciding parameters in star birth. In the case of **giant molecular clouds**, or GMCs, the conditions can give rise to both O- and B-type giant stars along with solar-type dwarf stars – whereas in **small molecular clouds** (SMCs) only solar-type stars will be formed, with none of the luminous B-type stars. An example of an SMC is the **Taurus Dark Cloud**, which lies just beyond the Pleiades.

An interesting aspect of open clusters is their distribution in the night sky. Surveys show that although well over a thousand clusters have been discovered, only a few are observed to be at distances greater than 25° above or below the galactic equator. Some parts of the sky are very rich in clusters – Cassiopeia and Puppis – and this is due to the absence of dust lying along these lines of sight, allowing us to see across the spiral plane of our Galaxy. Many of the clusters mentioned here actually lie in different spiral arms, and so as you observe them you are actually looking at different parts of the spiral structure of our Galaxy.

An open cluster presents a perfect opportunity for observing star colors (see Appendix 7). Many clusters, such as the ever and rightly popular Pleiades, are all a lovely steely blue color. On the other hand, Caldwell 10 in Cassiopeia has contrasting bluish stars along with a nice orange star. Other clusters have a solitary yellowish or ruddy orange star along with fainter white ones, such as Messier 6 in Scorpius. An often striking characteristic of open clusters is the apparent chains of stars that are seen. Many clusters have stars that arc across apparently empty voids, as in Messier 41 in Canis Major. Another word for a very small, loose group of stars is an **asterism**. In some cases there may only be five or six stars within the group.

---

[4] Astronomers call every element other than hydrogen and helium, metals.

Open clusters are groups of stars that are usually young and have an appreciable angular size and may have a few hundred components. **Globular clusters** are clusters that are very old, are compact and may contain up to a million stars, and in some cases even more. The stars that make up a globular cluster are called **Population II** stars. These are metal-poor stars and are usually to be found in a spherical distribution around the galactic center at a radius of about 200 light years. Furthermore, the number of globular clusters increases significantly the closer one gets to the galactic center. This means that particular constellations that are located in a direction towards the galactic bulge have a high concentration of globular clusters within them, such as Sagittarius and Scorpius.

The origin and evolution of a globular cluster are very different from an open or galactic cluster. All the stars in a globular cluster are very old, with the result that any star earlier than a G or F type star will have already left the main sequence and be moving toward the red giant stage of its life. In fact, new star formation no longer takes place within any globular clusters in our Galaxy, and they are believed to be the oldest structures in our Galaxy. In fact, the youngest of the globular clusters is still far older than the oldest open cluster. The origin of the globular clusters is a topic of fierce debate and research, with the current models predicting that the globular clusters may have been formed within the protogalaxy clouds that went to make up our Galaxy.

There are about 150 globular clusters ranging in size from 60 to 150 light years in diameter. They all lie at vast distances from the Sun, and are about 60,000 light years from the galactic plane. The nearest globular clusters, for example Caldwell 86 in Ara, lie at a distance of over 6000 light years, and thus the clusters are difficult objects for small telescopes.

Appendix 6

Double Stars

Double stars are stars that although they appear to be just one single star, will on observation with either binoculars or telescopes resolve themselves into two stars. Many stars may appear as double due to them lying in the same line of sight as seen from the Earth, and this can only be determined by measuring the spectra of the stars and calculating their red (or blue) shifts. Such stars are called **optical doubles**. It may well be that the two stars are separated in space by a vast distance. Some, however, are actually gravitationally bound and may orbit around each other, over a period of days or even years.

Many double stars cannot be resolved by even the largest telescopes, and are called **spectroscopic binaries**, the double component only being fully understood when the spectra are analyzed. Others are **eclipsing binaries**, such as Algol (β Persei), where one star moves during its orbit in front of its companion, thus brightening and dimming the light observed. A third type is the **astrometric binary,** such as Sirius (α Canis Majoris), where the companion star may only be detected by its influence on the motion of the primary star.

The brighter of the two stars is usually called the **primary** star, whilst the fainter is called the **secondary** or **companion**. This terminology is employed regardless of how massive either star is, or whether the brighter is in fact the less luminous of the two in reality, but just appears brighter as it may be closer.

Perhaps the most important terms used in double-star work are the **separation** and the **position angle** (PA). The separation is the angular distance between the two stars, usually in seconds of arc, and measured from the brighter star to the fainter. The position angle is the relative position of one star, usually the secondary, with respect to the primary, and is measured in degrees, with 0° at due north, 90° at due east, 180° due south, 270° at due west, and back to 0°. It is best described by an example (see Figure A.3), the double star γ Virginis, with components of magnitude 3.5 and 3.5, has a separation of 1.8 arcseconds, at a PA of 267° (epoch 2000.0). Note that the secondary star is the one always placed somewhere on the orbit, with the primary star at the center of the perpendicular lines. The separation and PA of any double star are constantly changing, and should be quoted for the year observed. When the period is very long, some stars will have no appreciable change in PA for several years; others, however, will change from year to year.

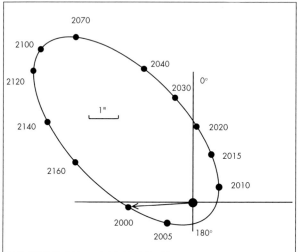

Figure A.3. The motion of γ Virginis.

Appendix 7

Star Colors

The most important factor which determines what the color of a star is, is you – the observer! It is purely a matter of both physiological and psychological influences. What one observer describes as a blue star, another may describe as a white star, or one may see an orange star, whilst another observes the same star as being yellow. It may even be that you will observe a star to have different color when using different telescopes or magnifications, and atmospheric conditions will certainly have a role to play. The important thing to remember is that whatever color you observe a star to have, then that is the color you should record.[5]

It may seem to a casual observer that the stars do not possess many bright colors, and only the brightest stars show any perceptible color: Betelgeuse can be seen to be red, and Capella is yellow, whilst Vega is blue, and Aldebaran has an orange tint, but beyond that most stars seem to be an overall white. To the naked eye, this is certainly the case, and it is only with some kind of optical equipment that the full range of star color becomes apparent.

But what is meant by the color of a star? A scientific description of a star's color is one that is based on the stellar classification, which in turn is dependent upon the chemical composition and temperature of a star. In addition, a term commonly used by astronomers is the color index. This is determined by observing a star through two filters, the B and the V filters, which correspond to wavelengths of 440 nm and 550 nm respectively, and measuring its brightness. Subtracting the two values obtained gives B – V, the **color index**. Usually, a blue star will have a color index that is negative, i.e. –0.3, orange-red stars could have a value greater than 0.0, and upwards to about 3.00 and greater for very red stars (M6 and greater). But as this is an observationally based book, the scientific description will not generally apply.

As mentioned above, red, yellow, orange and blue stars are fairly common, but are there stars which have, say, a purple tint, or blue, or violet, crimson, lemon, and the ever elusive green color? The answer is yes, but with the caution that it depends on how you describe the color. A glance at the astronomy books from the nineteenth century and beginning of the twentieth century will show you that star color was a hot topic, and descriptions such as amethyst (purple), cinerous (wood-ash tint), jacinth (pellucid orange), and smalt (deep blue), to name but a few, were used frequently. Indeed, the British Astronomical Association even had a section devoted to star colors. But today, observing and cataloging star color is just a pleasant pastime. Nevertheless, under good seeing conditions, with a dark sky, the keen-eyed observer will be able to see the faint tinted colors from deepest red to steely blue, with all the colors in between.

It is worth noting that several distinctly colored stars occur as part of a double-star system. The reason for this may be that although the color is difficult to see in an individual star, it may appear

---

[5] An interesting experiment is to observe a colored star first through one eye and then the other. You may be surprised by the result!

more intense when seen together with a contrasting color. Thus, in the section on double and triple stars, there are descriptions of many beautifully colored systems. For instance, the fainter of the two stars in η Cassiopeiae has a distinct purple tint, whilst in γ Andromadae and α Herculis, the fainter stars are most definitely green.

I have selected a few of the many fine astronomy and astrophysics books in print that I believe are amongst the best on offer. I do not expect you to buy, or even read them all, but check at your local library to see if they have some of them.

Star Atlases and Observing Guides

Amateur Astronomer's Handbook, J. Sidgwick, Pelham Books, London, UK, 1979
Burnham's Celestial Handbook, R. Burnham, Dover Books, New York, USA, 1978
Deep-Sky Companions: The Messier Objects, S. O'Meara, Cambridge University Press, Cambridge, UK, 1999
Millennium Star Atlas, R. Sinnott, M. Perryman, Sky Publishing, Massachusetts, USA, 1999
Norton's Star Atlas & Reference Handbook, I. Ridpath (Ed.), Longmans, Harlow, UK, 1999
Observing Handbook and Catalogue of Deep-Sky Objects, C. Luginbuhl, B. Skiff, Cambridge University Press, Cambridge, USA, 1990
Observing the Caldwell Objects, D. Ratledge, Springer-Verlag, London, UK, 2000
Sky Atlas 2000.0, W. Tirion, R. Sinnott, Sky Publishing & Cambridge University Press, Massachusetts, USA, 1999
The Night Sky Observer's Guide, Vols. I and II, G. Kepple, G. Sanner, Willman-Bell, Richmond, USA, 1999
Uranometria 2000.0 Volumes 1 & 2, Wil Tirion (Ed), Willmann-Bell, Virginia, USA, 2001

Astronomy and Astrophysics Books

Field Guide to the Deep Sky Objects, M. D. Inglis, Springer, London, UK, 2001
Galaxies and Galactic Structure, D. Elmegreen, Prentice Hall, New Jersey, USA, 1998
Introductory Astronomy & Astrophysics, M. Zeilik, S. Gregory, E. Smith, Saunders College Publishing, Philadelphia, USA, 1999

Observer's Guide to Stellar evolution, M. D. Inglis, Springer, London, UK, 2002
Stars, J. B. Kaler, Scientific American Library, New York, USA, 1998
Stars, Nebulae and the Interstellar Medium, C. Kitchin, Adam Hilger, Bristol, UK, 1987
The Milky Way, Bart & Priscilla Bok, Harvard Science Books, Massachusetts, USA, 1981
Voyages Through The Universe, A. Fraknoi, D. Morrison, S. Wolff, Saunders College Publishing, Philadelphia, USA, 2000

Magazines

Astronomy Now UK
 Sky & Telescope USA
 New Scientist UK
 Scientific American USA
 Science USA
 Nature UK

The first three magazines are aimed at a general audience and so are applicable to everyone; the last three are aimed at the well-informed lay person. In addition there are many research-level journals that can be found in university libraries and observatories.

Organizations

The Federation of Astronomical Societies, 10 Glan y Llyn, North Cornelly, Bridgend County Borough, CF33 4EF, Wales, UK
http://www.fedastro.demon.co.uk/

Society for Popular Astronomy, The SPA Secretary, 36 Fairway, Keyworth, Nottingham NG12 5DU, UK
http://www.popastro.com/

The American Association of Amateur Astronomers, P.O. Box 7981, Dallas, TX 75209-0981, USA
http://www.corvus.com

The Astronomical League
http://www.astroleague.org/

The British Astronomical Association, Burlington House, Piccadilly, London, W1V 9AG, UK.
http://www.ast.cam.ac.uk/~baa/

The Royal Astronomical Society, Burlington House, Piccadilly, London W1V 0NL, UK
http://www.ras.org.uk/membership.htm

The Webb Society
http://www.webbsociety.freeserve.co.uk/

International Dark-Sky Association, 3225 N. First Ave., Tucson, AZ 85719, USA.
http://www.darksky.org/

Campaign for Dark Skies, 38 The Vineries, Colehill, Wimborne, Dorset, BH21 2PX, UK.
http://www.dark-skies.freeserve.co.uk/

Appendix 9

The Greek Alphabet

The following is a quick reference guide to the Greek letters, used in the Bayer classification system. Each entry shows the uppercase letter, the lowercase letter, and the pronunciation.

A α	Alpha		H η	Eta		N ν	Nu		T τ	Tau
B β	Beta		Θ θ	Theta		Ξ ξ	Xi		Y υ	Upsilon
Γ γ	Gamma		I ι	Iota		O o	Omicron		Φ φ	Phi
Δ δ	Delta		K κ	Kappa		Π π	Pi		X χ	Chi
E ε	Epsilon		Λ λ	Lambda		P ρ	Rho		Ψ ψ	Psi
Z ζ	Zeta		M μ	Mu		Σ σ	Sigma		Ω ω	Omega

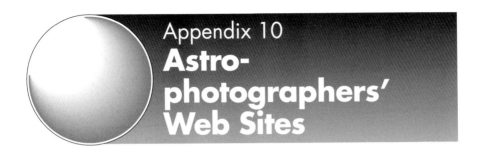

Appendix 10
Astro-photographers' Web Sites

Matt BenDaniel – http://starmatt.com/
Mario Cogo – www.intersoft.it/galaxlux
Bert Katzung – www.astronomy-images.com
Dr. Jens Lüdeman – http://www.ias-observatory.org/IAS/index-english.htm
Axel Mellinger – http://home.arcor-online.de/axel.mellinger/
Thor Olson – http://home.att.net/~nightscapes/photos/MilkyWayPanoramas/
Harald Straus [Astronomischer Arbeitskreis Salzkammergut] – http://www.astronomie.at/
Chuck Vaughn – http://www.aa6g.org/Astronomy/astrophotos.html
SEDS – http://www.seds.org/~spider/ngc/ngc.html

Index of Objects

The entry refers to its most familiar name and/or its main entry in the books. Page numbers are referred to by book number followed by page number.

Σ (Struve) 268 1-159
Σ (Struve) 270 1-159
Σ (Struve) 304 1-159
Σ (Struve) 336 1-159
Σ (Struve) 369 1-160
Σ (Struve) 390 1-125
Σ (Struve) 392 1-160
Σ (Struve) 484 1-128
Σ (Struve) 485 1-125, 1-128
Σ (Struve) 550 1-125
Σ (Struve) 559 1-184

Σ (Struve) 572 1-184
Σ (Struve) 698 1-171
Σ (Struve) 928 1-171
Σ (Struve) 929 1-171
Σ (Struve) 1108 1-190
Σ (Struve) 2303 1-37
Σ (Struve) 2306 1-41
Σ (Struve) 2325 1-44
Σ (Struve) 2373 1-42
Σ (Struve) 2404 1-47
Σ (Struve) 2470 1-88

Σ (Struve) 2474 1-88
Σ (Struve) 2576 1-73
Σ (Struve) 2587 1-47
Σ (Struve) 2634 1-60
Σ (Struve) 1999 2-162
Σ (Struve) 2445 1-65
Σ (Struve) 2876 1-92
Σ (Struve) 2894 1-92
Σ (Struve) 2922 1-93
Σ (Struve) 2940 1-92
Σ (Struve) 2942 1-92